Pathophysiology
PreTest® Self-Assessment and Review

Notice

Medicine is an ever-changing science. As new research and clinical experience broaden our knowledge, changes in treatment and drug therapy are required. The editors and the publisher of this work have checked with sources believed to be reliable in their efforts to provide information that is complete and generally in accord with the standards accepted at the time of publication. However, in view of the possibility of human error or changes in medical sciences, neither the editors nor the publisher nor any other party who has been involved in the preparation or publication of this work warrants that the information contained herein is in every respect accurate or complete, and they are not responsible for any errors or omissions or for the results obtained from use of such information. Readers are encouraged to confirm the information contained herein with other sources. For example and in particular, readers are advised to check the product information sheet included in the package of each drug they plan to administer to be certain that the information contained in this book is accurate and that changes have not been made in the recommended dose or in the contraindications for administration. This recommendation is of particular importance in connection with new or infrequently used drugs.

PreTest

Pathophysiology
PreTest® Self-Assessment and Review

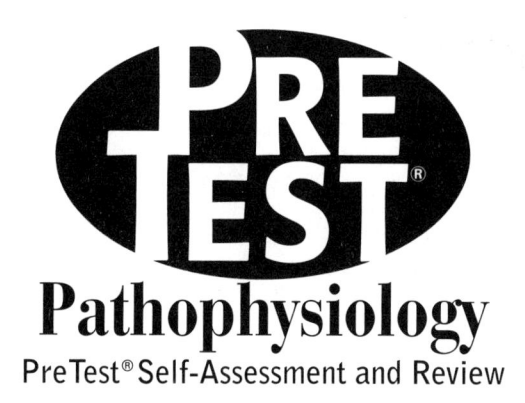

MAURICE A. MUFSON, M.D.
Professor and Chairman
Department of Medicine
Marshall University School of Medicine
Huntington, West Virginia

STUDENT REVIEWER

R. HENRY CAPPS, JR.
East Carolina University School of Medicine
Greenville, North Carolina

McGraw-Hill
Health Professions Division
PreTest® Series

NEW YORK   ST. LOUIS   SAN FRANCISCO   AUCKLAND
BOGOTÁ   CARACAS   LISBON   LONDON   MADRID
MEXICO CITY   MILAN   MONTREAL   NEW DELHI
SAN JUAN   SINGAPORE   SYDNEY   TOKYO   TORONTO

McGraw-Hill

*A Division of The **McGraw·Hill** Companies*

Pathophysiology: PreTest® Self-Assessment and Review

Copyright © 1999 by The McGraw-Hill Companies, Inc. All rights reserved. Printed in the United States of America. Except as permitted under the Copyright Act of 1976, no part of this publication may be reproduced or distributed in any form or by any means, or stored in a data base or retrieval system, without the prior written permission of the publisher.

1 2 3 4 5 6 7 8 9 0   DOCDOC   9 9 8

ISBN 0-07-052692-3

This book was set in Berkeley by V & M Graphics.
The editors were John J. Dolan and Susan R. Noujaim.
The production supervisor was Helene G. Landers.
The text designer was Jim Sullivan/RepoCat Graphics & Editorial Services.
The cover designer was Li Chen Chang / Pinpoint.
R.R. Donnelley & Sons was printer and binder.

This book is printed on acid-free paper.

The figure in the left column of page 62 and the figures on pages 63, 70 (bottom), and 73 are used courtesy of Lea & Febiger.

Acknowledgement

Mentors open doors for us in a way that only mentors can do. They promote our career and help us to see its direction. Their interest and understanding make the difference in the paths we take. Several mentors aided me throughout differing times of my career and I want to acknowledge them: Harold Heine, Ph.D., at Bucknell University, my first research mentor; the late Pinckney Jones Harman, Ph.D., at NYU School of Medicine; H. Sherwood Lawrence, M.D., also of NYU School of Medicine, who guided me into a career in Infectious Disease; Robert M. Chanock, M.D., at the National Institutes of Allergy and Infectious Diseases, who nurtured my research endeavors in virus diseases; Morton D. Bogdonoff, M.D., at the University of Illinois College of Medicine, who encouraged my becoming a Chair of a Department of Medicine; Erling Norrby, M.D., Ph.D., of the Karolinska Institute, Stockholm, Sweden, who opened his laboratory to me for my sabbatical and inspired me; and my wife, Deedee, who guides, encourages, nurtures, and inspires me in all my endeavors, and without whom my career would not have been the joy that it is.

Maurice A. Mufson, M.D., M.A.C.P.
Huntington, West Virginia

Contributors

Arif A. Aziz, M.D.
Assistant Professor of Medicine
Department of Medicine
Marshall University School of Medicine
Huntington, West Virginia

Anthony J. Bowdler, M.D.
Professor Emeritus
Department of Medicine
Marshall University School of Medicine
Huntington, West Virginia

Sachin T. Dave, M.D.
Assistant Professor of Medicine
Department of Medicine
Marshall University School of Medicine
Huntington, West Virginia

Henry K. Driscoll, M.D.
Professor of Medicine
Department of Medicine
Marshall University School of Medicine
Huntington, West Virginia

Martin L. Evers, M.D.
Assistant Professor of Medicine
Department of Medicine
Marshall University School of Medicine
Huntington, West Virginia

Lynne J. Goebel, M.D.
Assistant Professor of Medicine
Department of Medicine
Marshall University School of Medicine
Huntington, West Virginia

Susan H. Jackman, Ph.D.
Associate Professor
Department of Microbiology
Marshall University School of Medicine
Huntington, West Virginia

Imran T. Khawaja, M.D.
Assistant Professor of Medicine
Department of Medicine
Marshall University School of Medicine
Huntington, West Virginia

John W. Leidy, M.D.
Professor of Medicine
Department of Medicine
Marshall University School of Medicine
Huntington, West Virginia

Shirley M. Neitch, M.D.
Professor of Medicine
Department of Medicine
Marshall University School of Medicine
Huntington, West Virginia

Frank G. Renshaw, M.D.
Assistant Professor of Medicine
Department of Medicine
Marshall University School of Medicine
Huntington, West Virginia

W. Michael Skeens, M.D.
Assistant Professor of Medicine
Department of Medicine
Marshall University School of Medicine
Huntington, West Virginia

Marc A. Subik, M.D.
Assistant Professor of Medicine
Department of Medicine
Marshall University School of Medicine
Huntington, West Virginia

Paulette S. Wehner, M.D.
Assistant Professor of Medicine
Department of Medicine
Marshall University School of Medicine
Huntington, West Virginia

Jason L. Yap, M.D.
Assistant Professor of Medicine
Department of Medicine
Marshall University School of Medicine
Huntington, West Virginia

Contents

Contributors vii
Introduction xi

Immune System
Questions 1
Answers, Explanations, and References 6

Genetic Disease
Questions 13
Answers, Explanations, and References 20

Neoplasia and Blood Disorders
Questions 29
Answers, Explanations, and References 39

Infectious Disease
Questions 47
Answers, Explanations, and References 54

Cardiovascular
Questions 61
Answers, Explanations, and References 68

Pulmonary
Questions 79
Answers, Explanations, and References 83

Renal/Nephrology
Questions 89
Answers, Explanations, and References 95

GASTROENTEROLOGY

Questions 101
Answers, Explanations, and References 107

LIVER DISEASE

Questions 113
Answers, Explanations, and References 117

THYROID AND PITUITARY DISORDERS

Questions 121
Answers, Explanations, and References 128

FEMALE AND MALE REPRODUCTIVE TRACTS

Questions 135
Answers, Explanations, and References 142

NERVOUS SYSTEM

Questions 149
Answers, Explanations, and References 157

High-Yield Facts 165

Bibliography 171

INTRODUCTION

The questions we ask and the answers we verify about our patients and their disease lead to the correct diagnosis and treatment. Each patient represents a one-person clinical study. Clearly, questions form a part of the learning process that challenges and motivates us. They provide an approach to broadening our knowledge base and, in doing so, they can and must be a stimulus to additional reading. The questions in this book about pathophysiology are meant to do that. They cover the main points of each topic, and the answers include both a brief discussion of the information upon which the question is based and a source reference citation.

You might want to use the questions in this book in the following manner:

- Record your answer to the question before you read the correct answer.
- Look at the correct answer and the explanation.
- Read the source reference citation.

Such an approach can increase your understanding of the topic. The process of studying remains paramount, and not necessarily whether you know the correct answer to one question or many questions. It is important to read the source reference citation listed for each question, especially the questions for which you do not readily know the answer, to increase the scope of your knowledge, which after all is the goal of testing yourself on these questions.

The "High-Yield Facts," found at the end of the text, are condensed summaries and are provided to facilitate rapid review of topics in pathophysiology. It is anticipated that the reader will use the High-Yield Facts as a "memory jog" before proceeding through the questions.

Pathophysiology
PreTest® Self-Assessment and Review

IMMUNE SYSTEM

Questions

DIRECTIONS: Each item below contains a question or incomplete statement followed by suggested responses. Select the **one best** response or the **matching** response to each question.

Items 1–5

Match the immunoglobulin (Ig) class to the function listed:
a. IgA
b. IgG
c. IgM
d. IgE

1. The major immunoglobulin class in normal adult human serum

2. The predominant antibody found in a primary immune response

3. Found on the surface of mast cells

4. A major component of mucosal secretions

5. Can cross the placenta

6. Which of the following cells are important in an innate immune response to extracellular bacteria?
a. T cells
b. B cells
c. Neutrophils
d. Eosinophils

7. Which of the following is NOT a function of macrophages?
a. Present antigen to T cells
b. Produce cytokines
c. Secrete antibody
d. Are antibacterial

8. Compared with a healthy individual, lymph nodes from a patient with a deficiency in B cells would have
a. few or no primary follicles
b. enlarged germinal centers
c. reduced periarteriolar lymphocyte sheaths (PALS)
d. no paracortex

Pathophysiology

9. A newborn infected with group B streptococci would produce and secrete antibody of which of the following class(es)?
a. IgM only
b. IgG only
c. IgM and IgG
d. Neither IgM nor IgG

10. Eosinophils are associated with the defense against infections caused by
a. virus
b. intracellular bacteria
c. extracellular bacteria
d. parasites

11. To determine whether a fetus acquired an infection in utero, antigen-specific antibody of which of the following class should be measured?
a. IgA
b. IgM
c. IgG
d. IgD

12. During an immune response, antibodies are made against different structures (usually proteins) on an infectious agent. These structures are referred to as
a. adjuvants
b. allotypes
c. isotypes
d. epitopes

Items 13–16

Select the complement component that is associated with the activity listed:
a. C1
b. Factor B
c. C3b
d. C5a
e. C5b6789

13. Enhances phagocytosis of bacteria by opsonization

14. Mediates cytolysis

15. Chemoattractant for neutrophils

16. Binds to antibody to activate the classical pathway

17. A patient with a predisposition for infections by *Neisseria* bacteria may have a deficiency in
a. the membrane-attack complex (MAC) formation
b. classical pathway activation
c. C3
d. C1 inhibitor

18. Which of the following complement component deficiencies is associated with individuals with frequent pyogenic bacterial infections?
a. MAC
b. C1 inhibitor
c. C2
d. C3

19. A person with an abnormality in which of the following early complement components would most likely have the most serious clinical manifestations?
a. C1
b. C2
c. C3
d. Factor B

20. A 6-year-old boy has received a deep puncture wound while playing in his neighbor's yard. His records indicate that he has had the standard DPT (diphtheria, pertussis, tetanus) immunizations and a booster when he entered school. What is the most appropriate therapy for this child?
a. Tetanus toxoid
b. Tetanus antitoxin
c. Both toxoid and antitoxin at the same site
d. Toxoid and antitoxin at different sites
e. No treatment

21. Toxic shock syndrome toxin 1 is produced by some strains of *Staphylococcus aureus* and is thought to be responsible for the clinical manifestations of disease by this organism. This toxin is referred to as a superantigen because it can
a. activate T cells in an antigen-nonspecific manner
b. activate B cells without T cell help
c. become immunogenic when attached to a carrier protein
d. prolong the presence of antigen in a tissue

22. Antiviral activity of antibody includes all of the following EXCEPT
a. neutralization of extracellular virus
b. activation of natural killer (NK) cells
c. opsonization enhancing phagocytosis of virus
d. blocking of viral spread

23. Direct killing of virally infected cells is usually accomplished by
a. CD8-positive T cells
b. CD4-positive T-helper-1 cells
c. CD4-positive T-helper-2 cells
d. plasma cells

24. Mycobacterium tuberculosis results in an intracellular bacterial infection that provokes which of the following immune responses?
a. NK cytotoxic response
b. CD8-positive cytotoxic T cell response
c. T-helper-1 delayed-type hypersensitivity response
d. Complement mediated lysis of infected cells

25. During the course of an immune response to microbes, macrophages can cause damage to adjacent tissue by the release of all of the following EXCEPT
a. oxygen metabolites
b. hydrolytic enzymes
c. cytokines
d. histamine

26. During an immune response to pathogens in the intestine, the primary function of M cells along the Peyer's patches is to
a. transport antigen to lymphocytes
b. produce antigen-specific IgA antibody
c. present antigen to lymphocytes
d. secrete cytokines to "help" in antibody production

Items 27–31

Match the appropriate test for diagnosis to each of the applications listed:
a. Flow cytometry (FACS)
b. Enzyme-linked immunosorbent assay (ELISA)
c. Latex agglutination
d. Coombs' test
e. Mixed lymphocyte reaction

27. Determination of the titer of anti-hepatitis-B antibody

28. Detection of anti-Rh antibody in blood

29. Assessment of the level of CD4+ T cells in an HIV-infected patient

30. Evaluation of the degree of compatibility between donor and patient lymphocytes

31. Detection of group A streptococci from a throat swab

Items 32–38

Choose the appropriate hypersensitivity reaction associated with the condition described:
a. Type I: immediate
b. Type II: cytotoxic
c. Type III: immune complex
d. Type IV: cell mediated

32. Goodpasture's syndrome

33. Serum sickness

34. Tuberculin reaction

35. Poison ivy

36. Anaphylactic reaction after bee sting

37. Hemolytic disease of the newborn

38. Poststreptococcal glomerulonephritis

39. A patient with recurrent yeast infections and the incapacity to control viral infections may indicate a deficiency in
a. cellular immunity
b. complement
c. granulocytes
d. humoral immunity

40. Graft-versus-host (GVH) disease can be a complication of which of the following kinds of transplantation?
a. Kidney
b. Bone marrow
c. Liver
d. Skin

Items 41–45

Choose the cytokine that best fits the activity or function listed:
a. IFN-γ (interferon γ)
b. IL-2 (interleukin 2)
c. IL-4 (interleukin 4)
d. TNF-α (tumor necrosis factor α)
e. TGF-β (transforming growth factor β)

41. Promotes the proliferation of T and B lymphocytes

42. Promotes various biologic actions associated with inflammation

43. Antagonizes or suppresses many responses of lymphocytes

44. Promotes T-helper-2 (Th2) development and IgE synthesis

45. Activates macrophages and NK cells

Immune System

Answers

1. The answer is b. (*Murray, 5/e, p 93.*) IgG makes up about 85 percent of the immunoglobulin in adult serum.

2. The answer is c. (*Murray, 5/e, p 93.*) Most of the antibody produced in a primary immune response is IgM. As time passes or at a second encounter with the same antigen, isotype (class) switching can occur.

3. The answer is d. (*Murray, 5/e, p 93.*) IgE is found on the surface of mast cells and basophils. When antigen binds to the IgE, the mast cell releases various mediators involved in allergic reactions and antiparasitic defense.

4. The answer is a. (*Murray, 5/e, p 93.*) IgA is the predominant immunoglobulin class in mucosal secretions such as saliva, colostrum, bronchial, and genitourinary tract secretions.

5. The answer is b. (*Murray, 5/e, p 93.*) IgG can cross the placenta and confer passive immunity to the fetus and newborn.

6. The answer is c. (*Murray, 5/e, pp 110–111.*) Innate immunity involves antigen-nonspecific immune defense. Neutrophils circulate in the blood and can migrate into tissue to ingest and kill bacteria. Although T and B cells can augment the innate immune response, they become activated in an antigen-specific manner. Eosinophils, also part of an innate immune response, are important in parasitic, rather than bacterial, infections.

7. The answer is c. (*Murray, 5/e, p 82.*) B cells secrete antibody.

8. The answer is a. (*Murray, 5/e, p 86.*) The major cell type within follicles is the B cell; a germinal center is a follicle where cells are undergoing active proliferation. A deficiency in B cells would result in decreased size and number of follicles. The paracortex is predominately a T cell area. PALS is associated with the spleen.

9. The answer is a. (*Roitt, 5/e, p 168.*) A normal newborn can make IgM antibody in response to challenge with antigen. If IgG is detected in the newborn, it is most likely the result of placental transfer from the mother.

10. The answer is d. (*Murray, 5/e, p 88.*) Eosinophils are associated with parasitic infections. They localize near the parasite, degranulate, and release antiparasitic molecules. Eosinophils are not effective against intracellular bacteria or virus, which reside within host cells. Neutrophils are usually associated with extracellular bacterial infections.

11. The answer is b. (*Roitt, 5/e, p 168.*) The fetus and newborn infant can only produce measurable IgM antibody in response to infection. If IgG is detected, it is the result of an immune response by the mother and the antibody has crossed the placenta.

12. The answer is d. (*Murray, 5/e, pp 88 and 121.*) Usually an immunogen contains more than one molecule that can elicit an antibody response. These different molecules are called epitopes (or antigenic determinants) and are the structures with which antibodies react. Isotypes refer to the different classes of immunoglobulins (e.g., IgM, IgG, and IgA). Allotypes refer to isotypes that differ among individuals within a species. Adjuvants are substances that can enhance an immune response to antigen.

13. The answer is c. (*Murray, 5/e, pp 98–99.*) The cleavage component of C3, C3b, when bound to the surface of a cell, promotes the phagocytosis of that cell by a process referred to as opsonization.

14. The answer is e. (*Murray, 5/e, pp 98–99.*) Complement components C5b, 6, 7, 8, and 9 associate to generate the membrane-attack complex (MAC), which disrupts the integrity of the cell membrane on which it is formed.

15. The answer is d. (*Murray, 5/e, pp 98–99.*) Several complement cleavage products promote inflammatory responses: C3a, C4a, and C5a. They can also induce the degranulation of mast cells and so are also referred to as anaphylatoxins. C5a has the additional property whereby it is a neutrophil chemoattracting substance.

16. The answer is a. (*Murray, 5/e, pp 98–99.*) The classical pathway is initiated by antigen–antibody complexes. Binding of C1 to the complex activates the complement cascade.

8   Pathophysiology

17. The answer is a. (*Murray, 5/e, p 125.*) Patients with deficiencies in the components of the MAC would not be able to lyse foreign organisms. However, these individuals seem only to have a problem with disseminated neisserial infections.

18. The answer is d. (*Murray, 5/e, p 125.*) Individuals with C3 deficiency have recurrent serious pyogenic bacterial infections which can be fatal. The absence of C3 leads to the inability to generate the opsonin, C3b, which when deposited on the surface of the bacteria promotes phagocytosis. MAC deficiencies can lead to disseminated neisserial infections. A deficiency in C1 inhibitor is associated with hereditary angioneurotic edema (HANE). Individuals with C2 deficiency have a predisposition for immune-complex disease, such as systemic lupus erythematosus.

19. The answer is c. (*Murray, 5/e, p 98.*) C3 plays a central role in both the classical and alternate pathways. An abnormality in this component would disrupt both pathways. C1 and C2 are utilized only by the classical pathway; therefore, an abnormality in either one or both of these components would leave the alternate pathway intact. Likewise, a defect in factor B (a component of the alternate pathway) would still permit the activation of the classic pathway.

20. The answer is e. (*Murray, 5/e, pp 128–129.*) Since the boy received his booster within the last 2 years, his level of immunity should be adequate. If an individual has no history of immunization, both antitoxin (for temporary, fast protection) and toxoid (for future protection) should be given at different sites.

21. The answer is a. (*Murray, 5/e, p 156.*) A superantigen can activate T cells without binding to the T cell receptor in an antigen-specific manner. Therefore, the superantigen can stimulate a large number of T cells, which can result in massive cytokine release causing shock and tissue damage. An antigen that can activate B cells without T cell help is called a T-independent antigen. Haptens, usually small molecules, can become antigenic when attached to carrier proteins. Adjuvants can help in maintaining antigen at a tissue site.

22. The answer is b. (*Murray, 5/e, p 119.*) NK cells are activated by interferon α, β, or γ, and by interleukin 12.

23. The answer is a. (*Murray, 5/e, p 119.*) CD8-positive T cells are cytolytic T cells that can respond to viral peptides/MHC class I complexes on infected cells. CD4-positive T-helper-1 cells usually function by releasing cytokines that promote an inflammatory response. CD4-positive T-helper-2 cells produce cytokines important in generating antibody production. Plasma cells secrete antibody.

24. The answer is c. (*Murray, 5/e, p 114.*) Delayed-type hypersensitivity (DTH) responses are important in protection against intracellular bacteria. In this type of response, macrophages and other inflammatory processes are activated to kill the infected cell. NK cells, cytotoxic T cells, and complement do not appear to provide adequate protection against intracellular bacteria.

25. The answer is d. (*Murray, 5/e, pp 114 and 82.*) Histamine is released from mast cells and basophils.

26. The answer is a. (*Murray, 5/e, p 87.*) M cells deliver antigen to Peyer's patches, but they do not act as antigen-presenting cells to lymphocytes. Antibody is made by B cells within the Peyer's patch. T-helper-2 cells are the main source of cytokines functioning in helping B cells make and secrete antibody.

27. The answer is b. (*Murray, 5/e, pp 147 and 149.*) ELISA can be used to determine the relative antibody concentration to a specific antigen (titer); the assay can also be used to quantitate antibody.

28. The answer is d. (*Roitt, 5/e, pp 322–324.*) An indirect Coombs' test is used to detect circulating anti-Rh antibody: anti-Rh antibody reacts with Rh+ erythrocytes, causing agglutination of the erythrocytes. The direct Coombs' test tests for cell-bound anti-Rh antibody.

29. The answer is a. (*Murray, 5/e, pp 146 and 147.*) In flow cytometry, cells in suspension tagged with fluorescent-labeled antibody can be identified and quantitated.

30. The answer is e. (*Roitt, 5/e, pp 361–362.*) A mixed lymphocyte reaction assays the histocompatibility between two individuals. Donor cells are treated to prevent DNA synthesis and proliferation. The recipient's cells are mixed with the donor's cells. If the donor's cells express foreign MHC antigens, the recipient's lymphocytes will proliferate. Proliferation can be measured by the uptake of radioactive thymidine.

10   Pathophysiology

31. The answer is c. (*Murray, 5/e, pp 145 and 195.*) In a latex agglutination test, antigen-specific antibody is attached to latex beads. When the beads are mixed with a specimen containing antigen, the beads agglutinate, which can be detected visually.

32. The answer is b. (*Murray, 5/e, pp 123–124.*) In Goodpasture's syndrome, antibody forms to lung and kidney basement membranes, causing damage to the tissue.

33. The answer is c. (*Murray, 5/e, pp 123–125.*) Serum sickness results from the injection of serum made in nonhuman species into humans. Antibodies to the soluble nonhuman proteins are generated and immune complexes form. The complexes are trapped in capillaries and initiate an inflammatory response that damages tissue.

34. The answer is d. (*Murray, 5/e, p 125.*) Antigen injected intradermally into a previously sensitized individual elicits a delayed-type hypersensitivity response. This involves the recruitment of CD4+ T lymphocytes and macrophages to the site.

35. The answer is d. (*Murray, 5/e, p 125.*) Poison ivy is a contact dermatitis that draws CD4+ T lymphocytes and macrophages to the site of antigen.

36. The answer is a. (*Murray, 5/e, p 123.*) Anaphylaxis is a severe immediate hypersensitivity response. IgE, produced at the time of initial exposure to antigen (bee venom), binds to mast cells. Upon subsequent exposure, the antigen (bee venom) reacts with the mast cell-bound IgE, leading to the release of mediators from the mast cells. The mediators produce the symptoms associated with the anaphylactic reaction.

37. The answer is b. (*Murray, 5/e, pp 123–124.*) This is a cytotoxic hypersensitivity reaction in which IgG anti-Rh antibody, produced in a previous pregnancy, crosses the placenta and binds to Rh+ fetal red blood cells (RBCs). This triggers the classical complement pathway, leading to the lysis of the fetal red cells.

38. The answer is c. (*Murray, 5/e, pp 123–124 and 195.*) Complexes of bacterial antigen and antibody form and become trapped in the renal vasculature. Complement is activated and neutrophils are recruited to the site. During the process of removal of the immune complexes, tissue may be damaged.

39. The answer is a. (*Murray, 5/e, pp 126–127.*) Individuals with T cell deficiencies are susceptible to infections with microbes that reside within host cells (virus, *Mycobacterium* species, and fungi). Humoral immune deficiency or complement deficiency usually results in recurrent bacterial, rather than viral, infections. Granulocyte deficiency may also result in bacterial infections, as well as yeast infection. Since the statement indicates that the patient has problems with viral infections, the best answer is a deficiency in T cell immunity.

40. The answer is b. (*Murray, 5/e, p 123.*) GVH disease can develop in an immunosuppressed individual who receives immunocompetent donor cells. The donor cells must also be able to respond to histocompatibility antigens present on the recipient's cells that are NOT found on the donor cells. Bone marrow contains immunocompetent T cells; liver, kidney, and skin do not have a sufficient number of immunocompetent T cells to elicit GVH reactions.

41. The answer is b. (*Murray, 5/e, p 81.*) IL-2 acts on T cells to induce their progression through the cell cycle; it also acts as a growth factor for B cells.

42. The answer is d. (*Murray, 5/e, p 81.*) TNF-α activities depend partly on the quantity of cytokine produced. TNF-α is also associated with the production of IL-1 and IL-6. At low levels, it induces a local inflammatory effect by stimulating leukocyte recruitment. At moderate levels, it can have systemic effects, inducing fever and acute-phase protein synthesis within the liver. At high quantities, TNF-α (in conjunction with IL-1 and IL-6) can produce septic shock syndrome.

43. The answer is e. (*Murray, 5/e, p 81.*) TGF-β appears to be a signal that "turns off" inflammatory or immune responses.

44. The answer is c. (*Murray, 5/e, p 81.*) IL-4 promotes the development of the T-helper-2 subset of CD4+ T lymphocytes. It also is important for class switching to IgE.

45. The answer is a. (*Murray, 5/e, p 81.*) IFN-γ acts on macrophages to enhance killing of ingested microbes. It also stimulates the cytolytic activity of NK cells.

Genetic Disease

Questions

DIRECTIONS: Each item below contains a question or incomplete statement followed by suggested responses. Select the **one best** response to each question.

46. A patient who has the autosomal dominant gene for type I osteogenesis imperfecta has blue scleras and slightly reduced height, whereas his brother has multiple fractures and deformities. This is an example of
a. polymorphism
b. mutation
c. variable expressivity
d. fitness

47. Your patient has an autosomal dominantly inherited disease. The patient and his grandfather show evidence of disease, but the patient's father is asymptomatic. This is an example of
a. polymorphism
b. mutation
c. variable expressivity
d. reduced penetrance

48. Two patients have the same eye color. They have the same
a. HLA type
b. phenotype
c. haplotype
d. mutation

49. A patient has an X-linked disease. His three sisters do not have the disease. He most likely has
a. a mutant recessive gene on the X chromosome
b. a mutant dominant gene on the X chromosome
c. a mutant recessive gene on the Y chromosome
d. a mutant dominant gene on the Y chromosome

50. The fact that type IV osteogenesis imperfecta can be caused by defects on *COLIA* 1 and *COLIA* 2 is an example of
a. gonadal mosaicism
b. genetic heterogeneity
c. allelic heterogeneity
d. polymorphism

51. In genetics, fitness refers to
a. strong healthy chromosomes
b. genes fitting together on a single chromosome
c. absence of mutations
d. likelihood of reproduction by the individual with the mutant allele

52. The increased frequency of the recessive gene for sickle cell anemia in the African population is an example of
a. hypomorphism
b. hypermorphism
c. heterozygote advantage
d. phenotypic heterogeneity

53. Mutations that cause a gain in function of the mutated allele are
a. hypermorphic
b. neomorphic
c. amorphic
d. hypomorphic

54. When two copies of a mutant allele produce a phenotype more severe than one mutant and one normal copy, we have
a. dominant inheritance
b. recessive inheritance
c. semidominant inheritance
d. double dominant inheritance

55. Cystic fibrosis
a. occurs mainly in African Americans
b. causes endocrine problems with the pancreas gland
c. is an autosomal recessive disorder
d. is caused by loss of function mutations in a sodium channel

56. Your patient presents with multiple café-au-lait spots and neurofibromas. His father and mother do not have neurofibromas. This may be an example of
a. a new mutation
b. hypermorphism
c. a dominant negative mutation
d. antimorphism

57. A patient has muscular weakness. His parents and sister do not have weakness, but his mother's brother has weakness. You suspect Duchenne's muscular dystrophy. This is an example of
a. autosomal recessive inheritance
b. X-linked recessive inheritance
c. semidominant inheritance
d. autosomal dominant inheritance

58. Which of the following conditions is known to be multifactorial in etiology and not due to a single gene disorder?
a. Neurofibromatosis
b. Osteogenesis imperfecta
c. Atherosclerosis
d. Cystic fibrosis

59. A 13-year-old child with blue scleras, mildly short stature, no deformity, and a past history significant for 10 fractured bones most likely has
a. type I osteogenesis imperfecta
b. type II osteogenesis imperfecta
c. type III osteogenesis imperfecta
d. type IV osteogenesis imperfecta

60. An infant with multiple fractures, bony deformity, blue scleras, wormian bones in the skull, and beaded ribs died of respiratory difficulties. He most likely had
a. type I osteogenesis imperfecta
b. type II osteogenesis imperfecta
c. type III osteogenesis imperfecta
d. type IV osteogenesis imperfecta

61. Fragile X-associated mental retardation syndrome
a. is transmitted from father to son
b. affects females and males equally
c. in 20 percent of carrier males shows no sign
d. involves a nonrepetitive segment of DNA

62. Which is true regarding CpG islands?
a. They are less than 100 base pairs long.
b. They contain few sites for DNA methylation.
c. The island at Xq27.3 is normally unmethylated in male cells but methylated in one of the two X chromosomes in female cells.
d. Unmethylation of the CpG island in males correlates with expression of the fragile X-associated mental retardation syndrome.

63. A patient has fragile X-associated mental retardation syndrome phenotype. Diagnostic testing of 107 lymphocytes reveals repetition of $(5'\text{-CGG-}3')_n$ segment of DNA where $n > 200$, but variable in number. This variation in number is described as
a. genetic mosaicism
b. genetic anticipation
c. fitness
d. dosage compensation

64. The FMR1 protein
a. is expressed in fragile X-associated mental retardation
b. is normally found in brain and ovaries
c. is defective and not sufficient to cause the fragile X-associated mental retardation syndrome
d. is coded by the FMR1 gene, which has the $(5'\text{-CGG-}3')_n$ repeat segments

65. A premutation allele
a. if transmitted by a female expands to a full mutation with a likelihood proportionate to the size of the repeat segment
b. if transmitted by a male usually expands to a full mutation regardless of the length of the repeat sequence
c. causes a change in phenotype
d. is present in the "carrier" males

66. Fragile X, spinocerebellar ataxia 1, spinobulbar muscular atrophy, and Huntington's chorea
a. all involve synthesis of an altered protein with an expanded polyglutamine region
b. all involve failure to synthesize a protein
c. all involve amplification of an unstable triplet repeat
d. all involve mutation of the FMR1 gene

67. In the case of a dominant allele,
a. two copies of the allele are needed to produce the altered phenotype
b. one copy of the allele is sufficient to produce the phenotype
c. an offspring with one parent having a dominant allele has a 25 percent chance of inheriting the dominant allele
d. if two parents have a dominant allele, the offspring has a 50 percent chance of inheriting the dominant allele

68. All are true of human chromosomes EXCEPT
a. There are 23 pairs of allelic chromosomes in the parent cell.
b. A sperm contains only 23 chromosomes.
c. There are 22 pairs of autosomes and one pair of sex chromosomes.
d. A diploid cell has 23 chromosomes.

69. All are true of Down's syndrome EXCEPT
a. Patients have three copies of chromosome 21.
b. It may have been secondary to nondisjunction during meiosis.
c. It is an example of euploidy.
d. It may be inherited from a parent with a Robertsonian translocation.

70. Most children with Down's syndrome are
a. born to women under age 35
b. born to women over age 35
c. tall
d. underweight

Genetic Disease 17

71. Regarding life expectancy in Down's syndrome,
a. 90 percent survive to age 30 regardless of the presence of congenital heart disease
b. 60 percent survive to age 10, and 50 percent survive to age 30, if congenital heart disease is present
c. it is the same as for unaffected individuals if congenital heart disease is absent
d. 90 percent die of Alzheimer's disease by age 20

72. Which is true of Down's syndrome?
a. Fifty percent of cases are caused by an extra maternal chromosome.
b. Twenty-five percent of nondisjunctions occur in meiosis I.
c. Maternal and paternal nondisjunction events are associated with advanced maternal age.
d. It is impossible to tell whether the extra gene came from the mother or the father.

73. All are true of phenylketonuria (PKU) EXCEPT
a. It is successfully treated by diet.
b. It causes mental retardation if untreated.
c. Affected women may bear children with congenital anomalies regardless of infant genotype.
d. It is more common in African Americans than Northern Europeans.

74. Newborn screening for PKU
a. occurs at 7 days after birth
b. confirms the diagnosis in about 5 percent of those screened
c. has a false-negative rate of 1:70
d. shows a prevalence in the general population of 1:50,000

75. Regarding the treatment for PKU,
a. infants are fed a semisynthetic formula low in phenylalanine
b. breast-feeding is prohibited
c. infants are fed a diet totally devoid of phenylalanine
d. it can be discontinued at age 18

76. The neurologic deficits of PKU are due to
a. primarily the metabolites of phenylalanine
b. a direct effect of phenylalanine on energy production, protein synthesis, and neurotransmitter homeostasis
c. phenylalanine causing an increased transport of neutral amino acids across the blood–brain barrier
d. the increased action of phenylalanine hydroxylase

77. The relatively high incidence of PKU in the population could be due to all of the following EXCEPT
a. high mutation rate
b. founder effects
c. heterozygote advantage
d. low fitness of affected individuals

78. The different genetic forms of PKU illustrate two different pathophysiologic mechanisms by which inborn errors of metabolism cause disease. These are
a. end-product overproduction and substrate accumulation
b. end-product deficiency and substrate accumulation
c. end-product overproduction and substrate deficiency
d. end-product deficiency and substrate deficiency

79. The following is true regarding the centimorgan:
a. The human genome is composed of approximately 6000 centimorgans in recombination distance.
b. It is a measure of genetic distance that reflects the probability of a crossover between two loci during meiosis.
c. One centimorgan approximates a 5 percent chance of a crossover during meiosis.
d. The average chromosome contains about 500 centimorgans of genetic material.

80. The likelihood of two parents producing two offspring with identical chromosomes (other than by twinning) is
a. 1:540,000
b. 1:1,200,000
c. 1:5,800,000
d. 1:8,400,000

81. A mutation in which the base replacement changes the codon for one amino acid to another is called a
a. missense mutation
b. nonsense mutation
c. silent mutation
d. frameshift mutation

82. Southern blotting
a. was developed in the Southern United States
b. was named after E.M. Southern
c. is not useful for detecting gross rearrangements of DNA
d. cleaves DNA into large fragments

83. Polymerase chain reaction (PCR) technique for DNA amplification
a. is slow and cumbersome
b. can be used to detect nucleotide sequences of infectious agents
c. is not very specific
d. must be done on a fresh whole blood sample

84. Sickle cell anemia
a. is due to a single base change in the gene that codes for the β chain of hemoglobin
b. is an example of aneuploidy
c. involves substitution of glutamic acid for valine in the sixth amino acid position
d. is inherited as an X-linked recessive disorder

Genetic Disease 19

85. Anticipation refers to
a. waiting for a disease to manifest itself in an individual, such as in Huntington's chorea
b. worsening of a disease phenotype over generations within a family
c. birth of a fetus with positive prenatal tests for genetic abnormalities
d. expression of a premutation

86. A method to detect unique genomic DNA fragments from an individual starting with DNA from peripheral leukocytes is
a. northern blotting
b. southern blotting
c. eastern blotting
d. western blotting

87. All are true of autosomal recessive disorders EXCEPT
a. Parents are clinically normal.
b. Only siblings are affected.
c. Males and females are affected in equal proportions.
d. Consanguinity is not a contributing factor.

88. All are true of X-linked recessive disorders EXCEPT
a. Homozygous females do not occur.
b. Male offspring of a female carrier have a 50 percent chance of being affected.
c. All female offspring of affected males are carriers.
d. Unaffected males do not transmit the trait to any offspring.

89. The following is true regarding X inactivation:
a. It occurs late in embryonic development.
b. Each female is a mosaic with about half of her cells expressing the maternal X and half expressing the paternal X.
c. The nonfunctional X chromosome cannot be identified.
d. There is more methylation of DNA in the activated compared with the inactivated X chromosome.

90. Gene therapy has potential for cancer treatment. The strategies under consideration include all EXCEPT
a. alteration of cancer cells or other host cells to produce cytokines to alter host response to the malignancy
b. expression of antigens on cancer cells to induce a host immune response
c. insertion of tumor-suppressor genes into cancer cells to slow cell growth
d. introduction of drug-resistance genes into cancer cells

GENETIC DISEASE

Answers

46. The answer is c. *(McPhee, 2/e, p 6.)* The phenomenon of different phenotypes in individuals with the same genotype is known as variable expressivity. Polymorphism is an allele that is present in 1 percent or more of the population. Mutation refers to an event such as a nucleotide change, deletion, or insertion that produces a new allele. Fitness refers to the ability of an affected individual to reproduce.

47. The answer is d. *(McPhee, 2/e, p 5.)* Given a set of defined criteria, recognition of the condition in individuals known to carry the mutated gene is described as penetrance. Reduced penetrance is commonly seen in dominantly inherited conditions that have relatively high fitness, such as Huntington's disease or polycystic kidney disease.

48. The answer is b. *(McPhee, 2/e, p 5.)* A phenotype is any characteristic that can be described by an observer. HLA type refers to human leukocyte antigens that are coded by chromosome 6 and are especially important for transplant candidates. Haplotype refers to a set of closely linked alleles that are not easily separated by recombination. Mutation refers to an event such as a nucleotide change, deletion, or insertion that produces a new allele.

49. The answer is a. *(McPhee, 2/e, p 6.)* In males, only one copy of a mutant recessive gene on the X chromosome is sufficient to cause the genetic disease. Females, on the other hand, require two copies of the recessive gene. If this had been a dominantly inherited condition, chances are that at least one of his sisters would be affected.

50. The answer is b. *(McPhee, 2/e, p 13.)* Genetic heterogeneity is defined as a situation in which mutations of different genes produce similar or identical phenotypes. Gonadal mosaicism refers to mutation affecting some of the germ cells (sperm or eggs). Allelic heterogeneity refers to the state in which multiple alleles at a single locus can produce a disease phenotype or phenotypes.

51. The answer is d. (*McPhee, 2/e, p 17.*) A mutant allele that causes death in utero has low fitness. One in which the affected individual lives to reproductive age and allows the individual to pass on the mutant allele has high fitness.

52. The answer is c. (*McPhee, 2/e, p 18.*) When heterozygotes for a disease have a selective advantage compared with homozygous nonaffected individuals, this is called heterozygote advantage. This may account for the high incidence of certain mutant alleles in a population. One example is the high incidence of sickle cell anemia in persons of African descent. The heterozygous state for sickle cell anemia confers protection against malaria and offers a survival advantage.

53. The answer is a. (*McPhee, 2/e, p 6.*) Hypermorphic mutations cause a gain of function. Neomorphic mutations cause the acquisition of a new property. Amorphic mutations cause a complete loss of function. Hypomorphic mutations cause a partial loss of function.

54. The answer is c. (*McPhee, 2/e, p 7.*) Semidominant inheritance probably occurs in most dominantly inherited conditions, but homozygous mutant individuals are rarely observed. One example is when two people with achondroplasia have children. They have a 25 percent chance of producing a homozygous offspring. These children usually die in the perinatal period.

55. The answer is c. (*McPhee, 2/e, p 7.*) Cystic fibrosis, which is inherited as an autosomal recessive disorder, occurs most frequently in whites (1:2000) and can cause exocrine problems with the pancreas gland. The loss of function mutation responsible for cystic fibrosis is in a chloride channel.

56. The answer is a. (*McPhee, 2/e, p 7.*) About 50 percent of cases of neurofibromatosis are due to new mutations. Hypermorphism is a mutation that produces an increase in function. A dominant negative mutation gives rise to a protein that interferes with the function of the normal allele. One copy of the dominant negative allele has the same effect as two copies of the allele. This effect is called antimorphic.

22   Pathophysiology

57. The answer is b. *(McPhee, 2/e, p 7.)* Duchenne's muscular dystrophy is an X-linked recessive disorder. The patient's mother carries one copy of the recessive gene. It is not expressed in the patient's sisters, who may also carry one copy of the recessive gene.

58. The answer is c. *(McPhee, 2/e, p 8.)* Atherosclerosis is felt to be multifactorial. The effects of both genes and the environment play a role in its etiology. The other disorders listed are all single-gene disorders.

59. The answer is a. *(McPhee, 2/e, p 9.)* Type I osteogenesis imperfecta is mild. Type II is severe and usually lethal in the perinatal period. Type III is considered progressive and deforming. Type IV is deforming, but with normal scleras.

60. The answer is b. *(McPhee, 2/e, p 9.)* Wormian bones are isolated islands of mineralization in the skull. Type II osteogenesis imperfecta usually results in death in infancy.

61. The answer is c. *(McPhee, 2/e, p 14.)* X-linked disorders are passed on from mother to son. Males are affected more than females in this disorder. One-third of carrier females have a significant degree of mental retardation, and 20 percent of carrier males are nonpenetrant and manifest no signs of the disorder. The mutation involves a highly repetitive area of DNA.

62. The answer is c. *(McPhee, 2/e, p 16.)* CpG islands are several hundred base pairs long. They have many potential sites for DNA methylation. The CpG island at Xq27.3, the fragile site, is normally unmethylated in male cells, but methylated on one of the two X chromosomes in female cells. The CpG island at Xq27.3 is methylated in affected males and is methylated on both X chromosomes of affected females.

63. The answer is a. *(McPhee, 2/e, p 16.)* Patients with greater than 200 CGG repeat segments demonstrate the fragile X-associated mental retardation syndrome. The fact that, in an individual person, the actual number of repeats can vary from cell to cell is called genetic mosaicism. Genetic anticipation is demonstrated when a phenotype for a disease appears more severe in successive generations. Fitness is the ability of affected individuals to reach the reproductive age and transmit the mutation to offspring. Dosage compensation is the mechanism by which a difference in gene dosage between two cells is equalized. For example, in XX cells, one of the X chromosomes is inactivated, thereby providing a genetic dosage equal to an XY cell.

64. The answer is d. *(McPhee, 2/e, pp 16–17.)* The FMR1 protein is normally expressed in brain and testes. It is encoded by the FMR1 gene. Amplification of the CGG region to a repeat number greater than 200 causes methylation of the CpG island and prevents the FMR1 protein from being expressed. This defect is sufficient to cause the fragile X-associated mental retardation syndrome.

65. The answer is a. *(McPhee, 2/e, p 17.)* A premutation allele transmitted by a female expands to a full mutation with a likelihood proportionate to the length of the premutation. Premutation alleles with a repeat number between 52 and 60 rarely expand to a full mutation, whereas those with a repeat number greater than 90 nearly always expand. A premutation allele transmitted by a male rarely, if ever, expands to a full mutation regardless of the length of the repeat number. A premutation does not cause a change in phenotype. Carrier males have the mutation in their genes. About 20 percent of carrier males do not show evidence of the disease.

66. The answer is c. *(McPhee, 2/e, p 17.)* Spinocerebellar ataxia 1, spinobulbar muscular atrophy, and Huntington's disease are caused by expansion of a $(5'\text{-CAG-}3')_n$ repeat rather than the $(5'\text{-CFF-}3')_n$ repeat seen in fragile X. The first three diseases involve synthesis of a protein with an expanded polyglutamine region rather than failure to synthesize a normal protein as in fragile X-associated mental retardation syndrome. Only fragile X involves the FMR1 gene mutation.

67. The answer is b. *(Cecil, 20/e, p 134.)* One copy of the allele is sufficient to produce the phenotype in dominant inheritance. In recessive inheritance, two copies of the allele are necessary to produce the phenotype. An offspring with one parent having the dominant allele has a 50 percent chance of inheriting the allele. An offspring with both parents having the dominant allele has a 75 percent chance of inheriting the gene.

68. The answer is d. (*Cecil, 20/e, p 134.*) A diploid cell has 46 chromosomes or 23 pairs of chromosomes. Sperm and eggs contain half the number or 23 chromosomes. There are 22 pairs of autosomes and one pair of sex chromosomes.

69. The answer is c. (*McPhee, 2/e, p 18.*) Aneuploidy is a deviation from the normal number of chromosomes. People with Down's syndrome have three copies of chromosome 21. Euploidy is the normal number of chromosomes. When two homologous chromosomes fail to separate during meiosis, it is called nondisjunction. In a Robertsonian translocation, the parent has 46 chromosomes, but one chromosome 21 is fused via its centromere to another acrocentric chromosome.

70. The answer is a. (*McPhee, 2/e, pp 18–20.*) Screening programs for mothers over age 35 detect most Down's syndrome pregnancies in women of this age group. Because of this and the lower total numbers of births to women over age 35, most children with Down's syndrome are born to women under age 35. Individuals with Down's syndrome have statures 2 to 3 standard deviations below the average. Weight in affected individuals is mildly increased compared with the general population.

71. The answer is b. (*McPhee, 2/e, p 20.*) Survival depends on the presence or absence of congenital heart disease. Of those with heart disease, 60 percent survive to age 10, and 50 percent survive to age 30. Premature onset of Alzheimer's disease neuropathic changes is present in 100 percent of affected individuals by age 35. Frank dementia, however, is not detectable in all of these patients and may not play a large role in mortality.

72. The answer is c. (*McPhee, 2/e, p 20.*) Interestingly, both maternal and paternal nondisjunction events are associated with advanced maternal age. Molecular markers can be used to determine whether the extra chromosome came from the mother or the father. In studies using these markers, it was found that 75 percent of the extra chromosome 21 came from the mother and 75 percent of the nondisjunction events occurred during meiosis I.

73. The answer is d. (*McPhee, 2/e, p 24.*) PKU is more common among Yemenite Jews (1:5,000) and Northern Europeans (1:10,000) than among

African Americans (1:50,000). It is successfully treated by diet, but gene therapy is being investigated. Women who have the disease may bear children with congenital anomalies due to maternal transfer of elevated levels of PKU to the fetus.

74. The answer is c. *(McPhee, 2/e, p 24.)* Newborn screening for PKU is done at 24 to 72 h after birth. The diagnosis is confirmed in about 1 percent of those screened. The false-negative rate is 1:70. These children are picked up later when they exhibit developmental delays or seizures. PKU has a prevalence of 1:10,000.

75. The answer is a. *(McPhee, 2/e, p 24.)* Infants with PKU are fed a semisynthetic formula low in phenylalanine. This formula can be combined with regular breast-feeding. Since phenylalanine is an essential amino acid, infants do require a minimal amount of phenylalanine in their diet. Dietary restriction should be continued indefinitely, since even adults with hyperphenylalaninemia develop neuropsychological and cognitive deficits.

76. The answer is b. *(McPhee, 2/e, p 24.)* The neurologic defects of PKU are due primarily to phenylalanine itself and not the metabolites. Phenylalanine has a direct effect on energy production, protein synthesis, and neurotransmitter homeostasis. It causes decreased transport of neutral amino acids across the blood–brain barrier. Phenylalanine hydroxylase, which converts phenylalanine to tyrosine, is decreased in action.

77. The answer is d. *(McPhee, 2/e, p 26.)* There are several possible reasons why the incidence of PKU is so high. There may just be a high mutation rate of that gene. A founder effect occurs when the population was founded by a small ancestral group that contained a large number of carriers for the deleterious gene. Heterozygote advantage occurs when the heterozygotes have a survival advantage. High fitness would account for a high incidence of the disease, because affected individuals are reproducing and passing the mutations on to their offspring.

78. The answer is b. *(McPhee, 2/e, p 27.)* In PKU, there is a deficiency in the end products of phenylalanine metabolism, which are catecholamines and neurotransmitters. There is also a buildup of the substrate phenylalanine, which has its own adverse effects.

26 Pathophysiology

79. The answer is b. (*Fauci, 14/e, p 366.*) The centimorgan is a measure of genetic distance that reflects the probability of crossover between two loci during meiosis. The human genome is composed of approximately 3000 centimorgans in recombination distance. One centimorgan approximates a 1 percent chance of crossover during meiosis. The average chromosome contains about 130 centimorgans of genetic material.

80. The answer is d. (*Fauci, 14/e, p 367.*) The likelihood of two parents producing two offspring with identical chromosomes (other than by twinning) is 2^{23} or 1 in 8.4 million.

81. The answer is a. (*Fauci, 14/e, p 369.*) A missense mutation is one in which the base replacement changes the codon for one amino acid to another. A nonsense mutation is one in which the base replacement changes the codon to one of the termination codons. A silent mutation is one in which the base replacement does not lead to a change in the amino acid but only to the substitution of a different codon for the same amino acid. A frameshift mutation is one in which deletion or insertion of one or two bases occurs in a coding region and causes every codon distal to the mutation in the same gene to be read in the wrong triplet frame.

82. The answer is b. (*Fauci, 14/e, p 372.*) Southern blotting was named after E.M. Southern. It is a technique for analyzing DNA where DNA is cleaved into small fragments, fractionated by electrophoreses onto agarose gels, and processed so that specific sequences can be identified. Southern blotting can detect gross rearrangements in DNA and some point mutations.

83. The answer is b. (*Fauci, 14/e, p 372.*) PCR is a technique for amplifying DNA. It can be done on a fresh blood sample or isolated from dried blood filters, mouthwash, or even old tissue sections. One use is in detecting nucleotide sequences of infectious agents. PCR is a rapid technique that takes a single day. It is automated and extremely specific.

84. The answer is a. (*Fauci, 14/e, p 377.*) Sickle cell anemia is due to a single base change in the gene that codes for the β chain of hemoglobin. The change causes substitution of valine for glutamic acid in the sixth amino acid position in the protein sequence of the β chain. It is inherited as an autosomal recessive disorder. Aneuploidy refers to an abnormal number of chromosomes such as that seen in trisomy 21.

85. The answer is b. *(Fauci, 14/e, p 381.)* In genetics, anticipation refers to the worsening of a disease phenotype over generations within a family. The phenomenon of anticipation is due to increasing size of repeats in premutations that cause earlier onset of disease or a more severe phenotype. The mere presence of premutations does not cause changes in phenotype.

86. The answer is b. *(Fauci, 14/e, p 372.)* This is the definition of southern blotting. Northern blotting starts with RNA and can be used to detect the absence or presence of a particular mRNA. Western blotting or immunoblotting is a procedure designed to analyze protein antigens.

87. The answer is d. *(Fauci, 14/e, p 383.)* All are true except that consanguinity can most definitely increase the chances of inheriting an autosomal recessive disorder. The chance that two related individuals carry the same mutation passed on from a common ancestor is much higher than for unrelated individuals.

88. The answer is a. *(Fauci, 14/e, p 385.)* Homozygous females occur when an affected male mates with a carrier female.

89. The answer is b. *(Fauci, 14/e, p 386.)* X inactivation occurs early in embryonic development. Since the X chromosome that is selected for inactivation occurs independently and randomly in each cell, it would be expected that females are mosaic, with about half of their cells expressing the maternal X and half expressing the paternal X. The nonfunctional X chromosome can be identified as a condensed clump of chromatin called a Barr body. There is more methylation of DNA in the inactivated X chromosome compared with the activated X chromosome.

90. The answer is d. *(Fauci, 14/e, p 407.)* Research protocols are currently testing the alteration of cancer cells or other host cells to produce cytokines to alter host immune response to malignancy, and the expression of antigens on cancer cells to induce an immune response. The introduction of drug-resistance genes into normal cells could potentially protect normal cells against chemotherapy and allow higher-dose therapy to be given with reduced toxicity.

NEOPLASIA AND BLOOD DISORDERS

Questions

DIRECTIONS: Each item below contains a question or incomplete statement followed by suggested responses. Select the **one best** response or the **matching** response to each question.

91. A 48-year-old white woman has what she feels is a suspicious lump in her breast, but a mammogram does not reveal any suspicious lesions. Truthful statements concerning potential pitfalls in management and diagnosis include
  a. assuming that mammography is "diagnostic"
  b. assuming that a radiographic lesion seen on mammography is the same as a palpable lesion
  c. not letting a negative or nonsuspicious mammogram influence the judgment of whether a palpable mass needs to be biopsied
  d. assuming that a benign aspiration cytology is definitive

92. A 55-year-old man has lung cancer in the right middle lobe. Syndromes H associated with the lung cancer include
  a. hypocalcemia
  b. hypocortisolemia
  c. hyperphosphatemia
  d. acromegaly

93. A 30-year-old man has pain in the left scrotum. What is currently valid concerning types of tumor?
  a. Alpha fetoprotein (AFP) is only elevated in seminomas.
  b. The half-life of AFP is 24 to 36 h.
  c. Lactate dehydrogenase (LDH) is an important marker to follow tumor progression or regression.
  d. Human chorionic gonadotropin-β subunit (β-hCG) is only elevated in seminoma.

94. A 20-year-old man has headaches, blurred vision, and lateralizing neurologic deficits. An intracranial mass lesion is found. Microbial etiologies would include all of the following EXCEPT
  a. *Aspergillus*
  b. *Toxoplasma gondii*
  c. *Bacteroides* spp.
  d. *Escherichia coli*

Pathophysiology

95. An 18-year-old African-American woman comes to your medical clinic with enlarged, fixed, hardened, and matted lymph nodes in her neck. She denies weight loss, fevers, and night sweats. The etiology of this abnormality includes all EXCEPT

a. periodontal abscess
b. thyroid cysts
c. infection in a salivary gland
d. *Pseudomonas* spp.

96. Appropriate treatment for the aforementioned patient includes

a. a trial of antibiotics for 3 days
b. treat immediately with isonicotinoylhydrazine (INH) and rifampin
c. immediate lymph node biopsy
d. determination of VDRL, complete blood count (CBC), Epstein-Barr virus (EBV) titer, antinuclear antibody (ANA), and thyroid-stimulating hormone (TSH)

97. In a patient with multiple enlarged lymph nodes, which should be biopsied?

a. Groin nodes
b. Nodes in the axilla
c. Superficial cervical nodes
d. Periaortic lymph node with CT guidance

98. A 42-year-old white man has generalized lymphadenopathy, without fever or night sweats. His total white blood cell (WBC) count is elevated to 50,000, with small mature-appearing lymphocytes. A diagnosis of chronic lymphocytic leukemia (CLL) is ascertained. All of the following are typical characteristics of CLL EXCEPT

a. the presence of a "smudge cell"
b. focal or diffuse infiltration of the bone marrow core biopsy
c. hypergammaglobulin
d. autoimmune anemia and thrombocytopenia

99. The etiology of CLL is

a. due to radiation
b. due to a retrovirus
c. a familial disease
d. unknown

100. What cells are found in biopsies of patients with Hodgkin's disease?

a. Reed-Sternberg (RS) cells
b. RS precursors
c. Inflammatory cells
d. "Owl eye"-type cells
e. All of these

101. Which of the following is not a major example of inherited susceptibility to cancer?

a. Li-Fraumeni syndrome
b. Familial polyposis coli
c. Familial retinoblastosis
d. Peutz-Jeghers syndrome

102. Animate causes of cancers include all of the following EXCEPT
a. Epstein-Barr virus
b. papilloma virus
c. *Helicobacter pylori*
d. adenoviruses

103. Known cancer cytogenetic abnormalities include all EXCEPT
a. 9,22 translocation chronic myelogenous leukemia (CML)
b. 15,17 translocation acute myelogenous leukemia (AML) type M3
c. isochromosome 12 testes cancer
d. 2,6 translocation

104. Multiple endocrine neoplasia, type I (MEN I), syndrome is associated with what signs and symptoms?
a. Tumors of the anterior pituitary
b. Cancer of the parathyroid
c. Deletion of chromosome 11-Q13
d. All of these

105. Common tests used for early cancer detection include all of the following EXCEPT
a. mammography for breast cancer
b. digital rectal examination after 50 years of age for colorectal cancer
c. PAP smears for cervical cancer
d. CA-125 in ovarian cancer

106. Which of the following chemotherapeutic drugs is NOT associated with neutropenia?
a. Vincristine
b. Cyclophosphamide
c. Cytarabine (ara-C)
d. Carboplatin

107. The common chemotherapy-induced organ-specific toxicities include all of the following EXCEPT
a. cisplatin-induced nephrotoxicity
b. Velban (vinblastine sulfate) and neurotoxicity
c. bleomycin and lung toxicity
d. doxorubicin cardiotoxicity

108. Barriers to the management of cancer pain include all of the following EXCEPT
a. poor pain assessment
b. reluctance to report pain breakthroughs
c. poor narcotic availability
d. believing the patient's complaint of pain

109. Some basic principles of cancer pain management are the following EXCEPT
a. matching the analgesia to the degree of pain
b. titrating the analgesia regimen to the patient's response
c. anticipating and treating side effects
d. attempting to use injectable medications as much as possible

110. A 22-year-old man comes to the emergency room of your hospital because he has a diffuse, erythematous rash involving nearly all of his body. His total WBC count is greater than 100,000 cells/mm^3. He also complains of bone pain, severe irritability, weakness, fatigue, nausea and vomiting, constipation, photophobia, and polyuria. His electrocardiogram (ECG) shows shortening of the QT interval, prolongation of the PR interval, and nonspecific T-wave changes. The most likely cause of his symptoms is
a. hypercalcemia
b. hypocalcemia
c. hypophosphatemia
d. hyperkalemia

111. Treatment of the disorder of the aforementioned man would include all of the following EXCEPT
a. intravenous saline
b. intravenous diphosphonates
c. calcitonin
d. phosphate binders in the gastrointestinal tract

112. A 45-year-old white man with a limited small cell lung cancer presents to the emergency room of a local hospital and exhibits agitation and confusion, ataxia, nystagmus, peripheral sensory loss, and generalized weakness. The most likely etiology of this disorder is
a. hypercalcemia
b. paraneoplastic syndrome
c. cerebral vascular accident
d. myasthenia gravis

113. A 52-year-old white woman with breast cancer on adjuvant therapy presents with back pain that intensifies upon movement, and pain over the L-1 vertebral body when she coughs, and that radiates down her left lower extremity to her leg and foot. The most likely etiology of this disorder is
a. paraneoplastic disorder
b. trauma to the lumbar disc
c. muscular spasm of the intercostal muscles
d. possible spinal cord compression

114. In the aforementioned patient, the most effective initial treatment is
a. intravenous Decadron (dexamethasone)
b. orthopedic consultation
c. physical therapy techniques
d. intravenous narcotics

115. A 66-year-old white woman with a known history of small cell lung cancer comes to your office because of engorgement of her neck veins on the right side and over her chest wall. She also has cyanosis of the extremities, facial edema, and difficulty with her mentation. Her diagnosis is most likely

a. congestive heart failure
b. lymphatic obstruction of the upper body
c. superior vena cava syndrome
d. deep venous thrombosis

116. The multistep theory of carcinogenesis can be applied to what form of cancer?

a. Head and neck cancer
b. Breast cancer
c. Lung cancer
d. Colorectal cancer

117. Platelet production (thrombopoiesis) is affected by more than one cytokine. Which of the following sets appears to be the most important in platelet development?

a. IL-3, granulocyte colony-stimulating factor (G-CSF), and granulocyte-macrophage colony-stimulating factor (GM-CSF)
b. IL-4, IL-6, and thrombopoietin
c. Erythropoietin, thrombopoietin, and IL-6
d. IL-6 and thrombopoietin
e. IL-3, IL-4, and IL-6

118. Which of the following factors complexes with factor VIII, which is activated to factor VIIIa when released from the complex?

a. Factor XIII
b. High molecular weight kininogen
c. Von Willebrand factor (VWF)
d. Thromboplastin
e. Plasminogen

119. Which of the following factors is/are dependent on vitamin K for synthesis?

a. Factor II (prothrombin)
b. Factor VII (proconvertin)
c. Protein S
d. All of these
e. A and B only

120. Which of the following causes of anemia is NOT associated with larger than normal red cells (macrocytosis)?

a. Folic acid deficiency
b. Therapy with zidovudine (AZT)
c. Hypothyroidism
d. Autoimmune hemolytic disease
e. Anemia of chronic disorders

121. Which of the following causes of an elevated hemoglobin concentration in the blood is characterized by a LOW level of erythropoietin in the blood?
a. Chronic tobacco smoking
b. Dwelling at high altitudes, such as in the Andes
c. Erythrocytosis associated with renal tumors
d. Primary polycythemia (polycythemia vera)
e. Erythrocytosis secondary to chronic pulmonary insufficiency

122. Which of the following causes of an increase in the neutrophil count in the blood is NOT associated with an absolute increase in the number of circulating polymorphonuclear neutrophil leukocytes?
a. The neutrophil leukocytosis of acute infection
b. The leukocytosis associated with release of epinephrine
c. Neutrophil leukocytosis associated with tissue inflammation or necrosis
d. Chronic neutrophilic leukemia
e. None of these

123. In a typical case of iron deficiency, which of the following molecular forms that contains or can bind to iron, increases in the patient's serum?
a. Hemoglobin
b. Ferritin
c. Hemosiderin
d. Myoglobin
e. Transferrin

124. Which of the following disorders is/are associated with an increase in the platelet count above the normal range (thrombocytosis)?
a. Essential thrombocythemia
b. Iron-deficiency anemia
c. The postsplenectomy state
d. All of these
e. A and C only

125. Which of the following is NOT an accurate statement pertaining to pernicious anemia?
a. The condition results from an autoimmune process leading to failure of secretion of *intrinsic factor* by the gastric mucosa.
b. Ninety percent or more of patients have antibodies in the serum directed against parietal cell membrane proteins.
c. More than 50 percent of patients have antibodies in serum against intrinsic factor or the intrinsic factor–cobalamin complex.
d. Therapy with folic acid reverses the effects of the cobalamin deficiency of pernicious anemia.
e. Complete vitamin B_{12} deficiency develops slowly, because hepatic stores of the vitamin are normally adequate for several years.

126. Which of the following is NOT an accurate statement related to the thrombocytopenia induced by therapeutic heparin?
a. Between 10 and 15 percent of patients treated with heparin will develop thrombocytopenia.
b. The mechanism of the thrombocytopenia is immune in origin and is IgG antibody mediated.
c. In some cases, the platelets of the thrombocytopenic patient undergo platelet aggregation leading to platelet activation and thrombus formation.
d. Significant or severe hemorrhage is frequently experienced with the platelet count more than 10,000/mL.
e. Secondary endothelial cell injury can complicate the thrombocytopenia.

127. Which of the following statements related to circulating erythrocytes (red cells) is untrue?

a. The nuclei of the precursor cells to erythrocytes are extruded from their cells shortly before the red cells leave the bone marrow. Consequently, the presence of nucleated red cells in the peripheral blood should be regarded as abnormal and may indicate an underlying disease state.
b. In a thin blood smear stained with a Romanowsky stain (such as Wright's stain), the youngest cells (reticulocytes) can be recognized by a blue coloration (basophilia) as different from the majority of red cells present.
c. The average diameter of erythrocytes is about 8 μm, consequently they cannot flow through the smaller capillaries that have a diameter of 2 to 4 μm.
d. The protein of hemoglobin, which is the principal constituent of the red cell contents, is in tetrameric form, with two alpha and two beta subunits.
e. The iron atom of the hemoglobin molecule, which is essential to its function of carrying oxygen, is an intrinsic part of the heme complex attached to each subunit of protein.

128. Which of the following statements concerning the relationship of the neutrophil polymorphonuclear leukocyte (PMN) to infection with bacterial pathogens is incorrect?

a. The principal functions of the PMN are expressed in the tissues and not usually in the bloodstream, which is simply the transport path of the cells to their required site of action.
b. The cytoplasmic granules of PMN are essentially inert, but metabolically have only a vegetative role in maintaining cell viability.
c. When the cell numbers of PMN are reduced significantly, the probability of severe bacterial infection can be greatly increased.
d. The average duration of the period of circulation of PMN after entering the bloodstream is about 6 to 8 h.
e. An increasing need for PMN produced by infection is met in part by large numbers of immature cells (especially band cells) being released from the marrow pool into the bloodstream.

Neoplasia and Blood Disorders

Items 129–133

Each of the questions is related to the two disorders (A) and (B). For each numbered question, select a lettered answer:
a. If the item is associated with (A) only
b. If the item is associated with (B) only
c. If the item is associated with both (A) and (B)
d. If the item is associated with neither (A) nor (B)

(A) chronic lymphocytic leukemia (B-cell type)
(B) chronic myelocytic leukemia

129. The pathologic cells in a stained blood film have appearances very similar to those of normal mature white cells or their precursors.

130. On karyotyping, a well-defined chromosomal abnormality is pathognomonic of the condition.

131. The condition customarily terminates by transition to a blast cell phase with similarities to acute leukemia.

132. The principal complication of the condition is loss of resistance to infection.

133. Enlargement of the spleen is a significant physical finding in many patients.

Items 134–138

Each of the lettered options identifies a mechanism responsible for thrombocytopenia. For each of the numbered diseases or disorders given below, select the most appropriate lettered option. Each option may be used once, more than once, or not at all:
a. Decreased production
b. Maldistribution
c. Accelerated destruction

134. Vitamin B_{12} deficiency

135. Disseminated intravascular coagulation (DIC)

136. Immune (idiopathic) thrombocytopenic purpura (ITP)

137. Any increase in spleen size

138. Thrombotic thrombocytopenic purpura (TTP)

139. Which of the following factors contribute to abnormal clotting in blood vessels (thrombosis)?
a. Decreased blood flow
b. Blood vessel injury or inflammation
c. Changes in the intrinsic properties of blood
d. All of these
e. None of these

Items 140–142

For each of the questions, select from the list of lettered items the most appropriate description of the factor named in the question. Each lettered item may be used once, more than once, or not at all:

a. A protein cofactor that exposes the inactivation site of activated coagulation factor V, which can then be cleaved by a protease.
b. A factor capable of inhibiting the serine protease factors II, IX, X, XI and XII, a process accelerated by heparin or similar molecules.
c. A vitamin K–dependent factor, activated in the presence of thrombin to cleave activated factors V and VIII.

140. Protein C

141. Protein S

142. Antithrombin III (ATIII)

NEOPLASIA AND BLOOD DISORDERS

Answers

91. The answer is c. *(Fauci, 14/e, p 363.)* Any suspicious palpable mass should be biopsied despite a negative mammogram, which can happen from 10 to 15 percent of the time.

92. The answer is d. *(Fauci, 14/e, p 620.)* Ectopic acromegaly is a paraneoplastic disorder related to cancer, including that of the lung. Conversely, hypercalcemia, hypercortisolism, and hypophosphatemia also are disorders associated with lung cancer.

93. The answer is c. *(Fauci, 14/e, p 603.)* LDH is an important marker to follow in any germ cell tumor. AFP elevation is seen only in nonseminoma, whereas β-hCG is seen in both nonseminoma and seminoma. The half-life of AFP is 5 to 7 days.

94. The answer is D. *(Fauci, 14/e, p 2427.)* All brain masses are not malignancies. Of these organisms, only *E. coli* has not been described in a cranial mass, and the others are associated with immunodeficiency states.

95. The answer is d. *(Fauci, 14/e, p 346.)* *Pseudomonas* is not associated with nonmalignant lymphadenopathy.

96. The answer is d. *(Fauci, 14/e, p 346.)* See the answer to question 97.

97. The answer is c. *(Fauci, 14/e, pp 346 and 347.)* Cervical nodes are more likely to yield an etiology of disease than those in the axillae and supraclavicular regions and much more likely than nodes in the inguinal/femoral (groin) area.

98. The answer is c. *(Fauci, 14/e, p 699.)* Hypogammaglobulinemia is a hallmark of B-cell CLL.

99. The answer is d. *(Fauci, 14/e, p 699.)* This most common leukemia has no known etiology.

100. The answer is e. *(Fauci, 14/e, p 707.)* A plethora of cells is seen in the lymph node biopsies of Hodgkin's disease, mostly nonmalignant cells. The RS cell is thought to be the malignant cell in all Hodgkin's lymph node biopsies.

101. The answer is d. *(Fauci, 14/e, p 513.)* Li-Fraumeni syndrome, familial polyposis coli, and familial retinoblastosis are known genetic disorders associated with familial malignancies. Peutz-Jeghers syndrome, a familial disorder of multiple gastrointestinal polyps, rarely occurs as a familial cancer disorder.

102. The answer is d. *(Fauci, 14/e, pp 512 and 698.)* All except adenovirus have been associated with cancers in humans.

103. The answer is d. *(Fauci, 14/e, pp 605 and 685.)* All common chromosomal abnormalities except the 2,6 translocation.

104. The answer is d. *(Fauci, 14/e, p 2131.)*

105. The answer is d. *(Fauci, 14/e, pp 502–503.)*

106. The answer is a. *(Fauci, 14/e, pp 532–533.)* Vincristine causes peripheral neuropathy. It is used in combination because it does not cause neutropenia. The other chemotherapeutic drugs can cause profound myelosuppression, including neutropenia.

107. The answer is b. *(Fauci, 14/e, pp 532–533.)*

108. The answer is d. *(Fauci, 14/e, pp 497–498.)*

109. The answer is d. *(Fauci, 14/e, pp 497–498.)*

110. The answer is a. *(Fauci, 14/e, pp 618–619.)*

111. The answer is d. *(Fauci, 14/e, pp 618–619.)*

112. The answer is b. *(Fauci, 14/e, pp 618–622.)*

113. The answer is d. *(Fauci, 14/e, pp 628–629.)* Any back pain in a patient with a known history of carcinoma should be evaluated for the possibility of spinal cord disease.

114. The answer is a. *(Fauci, 14/e, pp 628–629.)* Intravenous Decadron is the choice for initial treatment in this situation because it will decrease swelling of the tumor mass.

115. The answer is c. *(Fauci, 14/e, p 627.)* The definitive diagnosis is superior vena cava syndrome until proven otherwise with scans.

116. The answer is d. *(Fauci, 14/e, p 517.)*

117. The answer is d. *(McPhee, 2/e, pp 99–100.)* Platelet production is stimulated by multiple cytokines, the most important being IL-6 and the peptide thrombopoietin. IL-3, IL-6, and GM-CSF also affect megakaryocytes, whereas erythropoietin and G-CSF are almost exclusively related to erythropoiesis and granulocyte production, respectively. IL-4 is also predominantly granulocyte related.

118. The answer is c. *(McPhee, 2/e, pp 104–105; Fauci, 14/e, pp 339–342.)* VWF complexes with factor VIII, which is activated by release from the complex to produce factor VIIIa. Together with factor IXa, calcium, and platelet phospholipid, factor VIIIa activates factor X. Factor XIII is activated to XIIIa by thrombin and improves the tensile properties of fibrin by chemical cross-linking. High molecular weight kininogen is the activator of factor XII. Thromboplastin is the lipid-rich protein material released on tissue injury that activates factor VII. Plasminogen is the precursor of plasmin, which is the serum protease that cleaves fibrin.

119. The answer is d. *(McPhee, 2/e, p 104; Fauci, 14/e, p 739.)* Factors II, VII, IX, and X are all dependent for synthesis on γ-carboxylase, a liver enzyme dependent on vitamin K. The two anticoagulant proteins S and C are also vitamin K dependent. Vitamin K is a necessary cofactor in the posttranslational synthesis of Y-carboxyglutamic acid groups in precursors of these factors.

42   Pathophysiology

120. The answer is e. (*McPhee, 2/e, p 105; Fauci, 14/e, pp 338 and 643.*) Macrocytic anemia results either from (1) abnormal nuclear maturation, known as nuclear–cytoplasmic asynchrony, resulting in megaloblastic changes in the bone marrow precursor cells. The principal causes are vitamin B_{12} and folic acid deficiency or drugs that interfere with DNA synthesis. (2) A high proportion of reticulocytes in the red cell population will also increase the average (mean) volume of red cells, as early red cells (reticulocytes) are larger than more mature red cells. This occurs when there is an active marrow proliferative response in compensation for active red cell destruction (hemolysis) or an active response to therapy for anemia such as vitamin B_{12}. The anemia of chronic disorders is usually normocytic in the early stages and may become microcytic later.

121. The answer is d. (*McPhee, 2/e, p 105; Fauci, 14/e, pp 679–680.*) The relative hypoxia of the tobacco smoker, the mountain dweller, and the patient with pulmonary insufficiency is a stimulus to erythropoietin production, which results in increased red cell production and circulating red cell mass. Likewise some tumors, including renal tumors, uterine myomata, and cerebellar hemangiomas may synthesize erythropoietin. Primary polycythemia is an abnormality of the bone marrow, leading to increased circulating red cell mass and feedback suppression of erythropoietin production.

122. The answer is b. (*McPhee, 2/e, p 106; Fauci, 14/e, pp 352 and 355.*) The leukocytosis associated with release of epinephrine, including conditions of stress (endogenous) and in therapy (exogenous), arises by demargination of neutrophils from the blood vessel walls. The apparent increase in the leukocyte count results from redistribution of the neutrophils, the marginated cells normally flowing close to the periphery of the blood vessels being relocated throughout the full volume of flowing blood. Leukocytosis associated with the stimulus of infection, inflammation, or necrosis is due to the proliferative stimulus increasing the true numbers of circulating cells. An absolute increase also occurs in myelogenous leukemia.

123. The answer is e. (*McPhee, 2/e, p 109; Fauci, 14/e, pp 337 and 640.*) Transferrin is the principal iron-binding protein present in plasma and carries the greater part of the iron in transport between the gut, storage sites and the bone marrow. In iron deficiency, the total iron-binding capacity of serum increases as the serum iron falls. The iron-binding capacity is principally dependent on the quantity of transferrin. Hemoglobin and myoglobin contain iron in the oxygen-carrying molecule heme. This binds oxygen reversibly, which permits transport by hemoglobin in red cells and storage by myoglobin in muscle.

124. The answer is d. *(McPhee, 2/e, p 107; Fauci, 14/e, p 683.)* High platelet counts occur in the myeloproliferative disorders, especially essential thrombocythemia; in this disorder, atrophy of the spleen is frequently a feature that worsens the thrombocytosis. High platelet counts occur in the hyposplenic states, including the postsplenectomy state, mainly because of redistribution of the excess platelets normally present in the spleen. Anemias, including iron-deficiency anemia and hemolytic anemias, may result in increased platelet counts, evidently reflecting the increased proliferative activity of the affected bone marrow.

125. The answer is d. *(McPhee, 2/e, pp 111–113; Fauci, 14/e, pp 653–659.)* Pernicious anemia results from a complex cascade of events that is autoimmune in origin. Antibodies against gastric parietal cell components and intrinsic factor are common, and antibody-generating B lymphocytes are found in the gastric mucosa. The signs of cobalamin deficiency are delayed by the liver storage of cobalamin, provided that the patient's intake has previously been normal. *Folic acid is not an adequate therapy for pernicious anemia:* partial reversal of megaloblastic blood cell changes may occur, but the neurologic changes resulting from vitamin B_{12} deficiency are unaffected. These may be more serious and disabling than the anemia, so therapy necessarily requires the use of cobalamin and not folic acid alone.

126. The answer is d. *(McPhee, 2/e, pp 117–119; Fauci, 14/e, pp 731 and 744.)* The incidence of thrombocytopenia is high in patients treated with heparin and should always remain in the mind of the clinician as a significant possibility. The pathogenesis involves binding of heparin to platelet factor 4 (PF4), and the released heparin–PF4 combination acts as an antigen provoking the production of an IgG antibody. The complex IgG–heparin–PF can bind to platelets by the platelet Fc receptor and lead to thrombocytopenia by destruction of the sensitized platelets in the spleen. However, the complex can also form bridges between platelets and induce aggregation with platelet activation and the potential for thrombus formation. Heparin-released PF4 molecules can also bind to heparin-like receptors on endothelial cells and induce cell injury. Heparin-induced thrombosis is sometimes known as the "white clot syndrome." The thrombocytopenia is usually mild and in the range of 50,000 to 100,000/mL.

127. The answer is c. *(McPhee, 2/e, pp 100–101.)* The presence of nucleated red cells in the peripheral blood is abnormal: it may be pathologic as in the leukoerythroblastosis that accompanies bone marrow infiltration, or with extramedullary erythropoiesis as in primary (agnogenic) myeloid metaplasia. Occasionally, it accompanies a brisk therapeutic correction of anemia. The earliest red cells (reticulocytes) still contain some ribosomes, mitochondria, and RNA and appear faintly basophilic (blue) in a Wright's stained blood smear. Hemoglobin is a tetrameric protein, and each subunit is associated with a heme complex containing the iron atom of the molecule, which is related to the locus of the carried oxygen atoms. The red cell is, in fact, normally a highly flexible body, capable of considerable modification of shape in traversing small capillaries. The flexibility may be compromised by increased intracellular viscosity or rigidity of the cell membrane, as in various hemolytic anemias.

128. The answer is b. *(McPhee, 2/e, p 101; Fauci, 14/e, p 351.)* The neutrophils (PMNs) are the predominant form of the WBCs, but their major function is in the tissues where they accumulate at sites of infection or inflammation, after transient passage through the bloodstream. Decreased available numbers (neutropenia) can result in a high incidence of bacterial infections. The granules contain enzymes with bactericidal properties, such as myeloperoxidase and NADPH oxidase.

129–133. The answers are b, b, b, a, c. *(McPhee, 2/e, p 106; Fauci, 14/e, pp 691–694 and 697–699.)* The chronic lymphocytic and myelocytic leukemias are characterized by proliferation of lymphoid and myeloid cells, which are usually present in excessive numbers in the peripheral blood: their appearances in the peripheral blood smear are usually close to those of the related normal forms, although other properties of the cells may be abnormal. Precursor forms are often prominent in myeloid leukemias. A characteristic chromosomal translocation t(9:22) results in the Philadelphia chromosome: although not exclusively restricted to chronic myelocytic leukemia, its presence in a chronic leukemia makes the diagnosis highly probable. The myelocytic leukemia commonly terminates following transition to an accelerated phase with transformation of the principal malignant cell to a blastlike form. The lymphocytic leukemia involves impaired antibody production and other immune functions, resulting in susceptibility to severe infections. Both forms of leukemia lead to increasing accumulation of malignant cells, leading to organ enlargement, especially of the spleen, liver, lymph nodes, and bone marrow.

134–138. The answers are a, c, c, b, c. *(McPhee, 2/e, p 107; Fauci, 14/e, pp 344 and 730–731.)* Thrombocytopenia is most commonly produced by processes that reduce the survival of circulating platelets significantly below the normal average life span of 10 days. In disseminated intravascular coagulation, activation of the coagulation sequence by infection, release of thromboplastins from malignant cells, hypoxia, or hemorrhage leads to a consumption coagulopathy that depletes the components of coagulation mechanisms, including the platelets. In ITP, autoimmune antibodies attack the platelet surface and initiate phagocytosis by attachment to the receptors of macrophages, especially in the spleen. A proportion, often about 10 percent, of the circulating platelets is normally present in a platelet pool in a normal spleen. With splenic enlargement (splenomegaly), the pool accommodates a higher proportion of the total and reduces the platelet count.

139. The answer is D. *(Fauci, 14/e, p 339.)* Each of these factors can contribute significantly to the potential for thrombosis; these are the components of Virchow's Triad *(McPhee, 2/e, p 120.)* Decreased flow enables the easier aggregation of formed blood elements, and blood vessel injury provides a damaged endothelial surface and the potential for platelet activation, whereas abnormal intrinsic properties of blood may include increased viscosity, which impairs blood flow, and alterations in the specific component factors, which inhibit coagulation.

140–142. The answers are c, a, b. *(McPhee, 2/e, pp 120–122; Fauci, 14/e, pp 341–342.)* Control of the coagulation system depends in large part on the activity of negative control factors that impede the excessive development of active coagulation. Protein C is an anticoagulation factor requiring vitamin K for its synthesis. Thrombin generated by the coagulation process and modified by thrombomodulin activates protein C, which cleaves factors Va and VIIIa and inhibits coagulation. Platelet phospholipid, calcium, and a cofactor, protein S, are also required. Antithrombin III (ATIII) is also an inhibitor of coagulation not only of thrombin, but of activated IX, X, XI, and XII. It acts by binding to the factor and not by enzymatic action. Its activity is very dependent on its accelerator cofactor, heparin.

INFECTIOUS DISEASE

Questions

DIRECTIONS: Each item below contains a question or incomplete statement followed by suggested responses. Select the **one best** response, the **matching** response, or the **true/false** response to each question.

143. Acute bacterial infections of the bone characteristically show all of the following EXCEPT
a. bacteria
b. polymorphonuclear leukocytes
c. congested blood vessels
d. thrombosed blood vessels
e. granulation tissue

144. In acute hematogenous and contiguous focus osteomyelitis, which of the following organisms accounts for at least 50 percent of both of these infections?
a. Group A streptococci
b. Group B streptococci
c. *Mycoplasma*
d. *Staphylococcus aureus*
e. *Pseudomonas aeruginosa*

145. Bacteria can infect the skin through accidental or deliberate breaks in it or through the hair follicle. Which bacteria causes one of several differing infections of the skin, including necrotizing fasciitis, erysipelas, impetigo contagiosa, and necrotizing myositis?
a. *Clostridium* spp.
b. *Streptococcus pyogenes*
c. *Staphylococcus aureus*
d. Anaerobic bacteria
e. *Pseudomonas aeruginosa*

146. Which of the following does not contribute to injection drug users becoming infected?
a. Unsterile injection technique
b. Immune defects induced by drug use
c. Contaminated needles and syringes
d. Non-use of antibiotics
e. Poor dental hygiene

147. Infective endocarditis frequently occurs in injection drug users. The valve most often involved is
a. mitral
b. aortic
c. tricuspid
d. pulmonic

148. Which of the following types of bites is more likely to become infected?
a. Human
b. Dog
c. Cat
d. Rat

149. Among nosocomial (hospital acquired) infections, which occurs most commonly and also causes the least sequelae?
a. Pneumonia
b. Urinary tract
c. Surgical wound
d. Bacteremia

150. Gram-negative and gram-positive bacteria each possess which of the following structures?
a. Peptidoglycan
b. Lipopolysaccharide
c. Matrix protein
d. Pili
e. None of these

151. Which organism is not a likely cause of left-sided infective endocarditis?
a. *Clostridium* spp.
b. *Staphylococcus aureus*
c. Viridans streptococcus
d. *Enterococcus*
e. None of these

152. The pathogenesis of bacterial meningitis involves all of the following EXCEPT
a. colonization of the nasal mucosa
b. access to the blood
c. antigen–antibody interaction
d. crossing of the blood–brain barrier
e. replication in the central nervous system

153. Commonly which of the following microorganisms is not likely to cause meningitis in children under 2 months of age?
a. *Neisseria meningitidis*
b. *Escherichia coli*
c. Group B streptococci
d. *Listeria monocytogenes*
e. None of these

Infectious Disease 49

154. Which of the following microorganisms is the most common cause of community acquired pneumonia?
a. Mycoplasma pneumoniae
b. Streptococcus pneumoniae
c. Staphylococcus aureus
d. Haemophilus influenzae
e. Legionella spp.

155. Which of the following microorganisms is not a likely cause of pneumonia among persons with human immunodeficiency virus (HIV) infection and AIDS?
a. Mycoplasma pneumoniae
b. Streptococcus pneumoniae
c. Pneumocystis carinii
d. Haemophilus influenzae
e. Mycobacterium tuberculosis

156. Which of the following bacteria is deposited directly into the lower airways?
a. Mycoplasma pneumoniae
b. Streptococcus pneumoniae
c. Pneumocystis carinii
d. Haemophilus influenzae
e. Mycobacterium tuberculosis

Items 157–161

Match the risk factor to the microorganism likely to cause pneumonia in the presence of the risk factor:
a. Legionella spp.
b. Pneumocystis carinii
c. Chlamydia psittaci
d. Klebsiella pneumoniae
e. Moraxella catarrhalis

157. Person with HIV and AIDS

158. Alcoholism

159. Exposure to an animal

160. Chronic lung disease

161. Transplant recipient

162. Which of the following groups of microorganisms more commonly causes diarrhea in the United States?
a. Bacteria
b. Fungi
c. Protozoa
d. Viruses

163. The single most important bacteria that causes diarrhea is
a. Helicobacter pylori
b. Staphylococcus aureus
c. Salmonella spp.
d. Shigella spp.
e. Escherichia coli

164. The pathogenesis of bacterial diarrhea involves all of the following EXCEPT
a. antigen–antibody aggregation
b. cytotoxin
c. enterotoxin
d. invasion of mucosal wall

165. The endogenous mediators of sepsis include all of the following EXCEPT
a. cytokines
b. endorphins
c. arachidonic acid metabolites
d. complement C5a
e. exotoxin

166. The principal CD lymphocyte affected in HIV and AIDS is
a. CD28
b. CD27
c. CD8
d. CD4
e. CD3

167. The destruction of CD4 lymphocytes in HIV and AIDS involves all of the following EXCEPT
a. direct viral destruction of CD4 lymphocytes
b. apoptosis
c. autoimmunity
d. syncytium formation
e. bone marrow stimulation

168. In addition to the CD4 T lymphocytes, all of the following can contribute to the pathogenesis of HIV and AIDS EXCEPT
a. eosinophils
b. macrophages
c. lymph nodes
d. monocytes
e. mononuclear cells

169. Typically HIV progresses from the onset of infection to evidence of immunosuppression over what period of time?
a. 1 to 3 months
b. 1 to 3 years
c. 4 to 5 years
d. 5 to 10 years
e. 11 to 15 years

Items 170–177

Characteristically, the immunosuppression of HIV and AIDS is heralded by the development of complicating illnesses, often life threatening. Match the information listed in A through E with the complicating illness:

a. *Candida albicans*
b. Cytomegalovirus
c. *Cryptococcus neoformans*
d. Herpesvirus varicella
e. *Cryptosporidium*
f. Kaposi's sarcoma
g. *Pneumocystis carinii*
h. *Mycobacterium tuberculosis*
i. *Mycobacterium avium–intracellulare* complex
j. *Toxoplasma gondii*

170. Lung infection, the most common opportunistic infection in HIV

171. Esophagitis, with substernal pain and dysphagia

172. Biliary tract infection, including sclerosing cholangitis

173. Painful crusted lesions following the path of an intracostal nerve

174. Headache, seizures, altered mental status, space-occupying lesion

175. Localized skin lesion or disseminated visceral lesion characterized by a mixed cell population that includes vascular endothelial cells

176. Retinitis, with patient complaints of blind spots, and hemorrhages and exudates seen on examination of the retina

177. Wasting disease and cachexia

178. Rickettsial disease is all of the following EXCEPT

a. gram-negative coccobacilli
b. intracellular organisms
c. transmitted by actual insect bite
d. grow in eukaryotic cells

179. Clinical manifestation of leptospirosis includes

a. jaundice and renal dysfunction
b. conjunctivitis
c. rash
d. headache
e. all of these

180. Rabies vaccination is indicated in all of the following EXCEPT

a. fox bite
b. health care workers in contact with animals (veterinarians)
c. neighborhood dog bite from an unvaccinated dog in a low endemic area
d. bat bite

181. Differential diagnosis of intestinal amebiasis includes disease caused by
a. Shigella
b. Salmonella
c. Campylobacter
d. inflammatory bowel disease
e. all of these

182. Primary syphilis is manifested by all of the following EXCEPT
a. single painless papule
b. presence of enlarged inguinal lymph nodes
c. rectal chancre with perirectal lymphadenopathy
d. generalized lymphadenopathy

183. INH prophylaxis is recommended for the treatment of tuberculosis in all EXCEPT
a. an asymptomatic HIV patient with a PPD of 7 mm
b. when the PPD ≥ 5 mm in a person with fibrotic lesion on chest x-ray
c. when the PPD ≥ 10 mm in a recently infected individual
d. an asymptomatic 34-year-old internist with a PPD >10 mm
e. an asymptomatic 40-year-old housewife with a PPD = 10 mm

184. Varicella pneumonia is a complication of chicken pox and (answer true or false for each)
a. develops more commonly in children
b. occurs late in the illness (after 7 to 10 days)
c. presents with hemoptysis and fever
d. has x-ray evidence of nodular infiltrates
e. resolution parallels improvement of rash

185. Which is the best way to diagnose cytomegalovirus (CMV)? Answer true or false for each:
a. Clinically with fever and rash
b. Isolation of the virus from urine
c. Persistently elevated viral titer

186. Antimalarial drugs include
a. quinidine
b. mefloquine
c. primaquine
d. doxycycline
e. chloroquine
f. all of these

187. Which of the following statements regarding cat scratch disease are correct? Answer true or false for each:
a. Sixty percent of cases occur in adults.
b. Fever and rash occur within 3 to 5 days of scratch in almost all cases.
c. Anorexia and malaise are common.
d. Lymphadenopathy may be mistaken for lymphoma.
e. Encephalitis, seizures, and coma can occur.

188. Which of the following statements regarding toxic shock syndrome (TSS) are correct? Answer true or false for each:
a. It can be confused with Stevens-Johnson syndrome.
b. It occurs only with the use of tampons.
c. It is characterized by high fever and erythrodermas.
d. Desquamation occurs typically on palms and sores.
e. It is caused by the exoproteins produced by *Staphylococcus aureus*.

INFECTIOUS DISEASE

Answers

143. The answer is e. *(Fauci, 14/e, p 824.)* Acute osteomyelitis does not show granulation tissue that occurs in chronic osteomyelitis. Additionally, necrotic bone, presence of granulation and fibrous tissues, very few bacteria, and the absence of living osteocytes characterize chronic osteomyelitis.

144. The answer is d. *(Fauci, 14/e, pp 824–825.)* In acute hematogenous osteomyelitis, *S. aureus* accounts for about 50 percent of infections, likely as the single organism. In contiguous focus osteomyelitis, *S. aureus* also occurs in more than 50 percent of cases, except it likely occurs together with other organisms as a polymicrobial infection.

145. The answer is b. *(Fauci, 14/e, p 828.)* *Streptococcus pyogenes* causes these differing skin infections because they infect the dermis and can spread laterally by the lymphatics to deeper and superficial areas. *Pseudomonas aeruginosa* causes hot-tub folliculitis especially in tubs that fail to maintain high water temperature, for example, between 37° and 40°C, and sufficient chlorination. *Staphylococcus aureus* causes bullous impetigo, furunculosis, and pyomyositis. *Clostridium* species causes gas gangrene.

146. The answer is d. *(Fauci, 14/e, p 831.)* Usually, intermittent antibiotic usage by injection drug users alters normal microbial flora, leading to increased risk of infection, and the nonuse of antibiotics would prevent it. All other factors contribute to the increased risk of infection in injection drug users.

147. The answer is c. *(Fauci, 14/e, p 832.)* The tricuspid valve is more frequently involved than the other valves, perhaps because of its nearness to the injection sites. In an individual patient, however, any of the heart valves may become infected. Left-sided infective endocarditis usually develops when underlying valvular disease exists.

Infectious Disease   Answers   55

148. The answer is a. *(Fauci, 14/e, p 837.)* Of the common bites, human bites more often become infected than animal bites. They occur as occlusional injuries: actual biting injuries and clenched-fist injuries sustained by striking the teeth of another individual. Infections caused by human bites reflect the multiple microorganisms that can be present in the mouth.

149. The answer is b. *(Fauci, 14/e, p 848.)* Urinary tract infections acquired in the hospital develop more commonly than any other nosocomial infections and have the fewest severe sequelae. The other nosocomial infections, especially pneumonia and bacteremia, are life threatening, are much more difficult to treat, and can lead to severe sequelae.

150. The answer is a. *(Fauci, 14/e, p 853.)* Gram-positive and gram-negative bacteria each have a peptidoglycan layer; all other structures are features of gram-negative bacteria only.

151. The answer is a. *(McPhee, 2/e, p 57.)* The most common bacteria causing infective endocarditis of the left side of the heart are *S. aureus*, viridans streptococcus, and *Enterococcus*. *S. aureus* commonly infects the right side of the heart in injection drug users.

152. The answer is c. *(McPhee, 2/e, p 61.)* The pathogenesis of meningitis does not rely on an antigen–antibody reaction, but it does rely on the other features. Usually, colonization of the nasopharynx is the first event in an infection that progresses to meningitis. Normally, IgA present on the nasal mucosa inhibits bacterial attachment, but many bacteria produce an IgA protease that cleaves IgA and permits their attachment. Bacteria then invade the vascular compartment and, providing they can overcome various host defense mechanisms, they infect the central nervous system and cause meningitis. The bacteria capsule promotes initiation of infection, because it inhibits phagocytosis by polymorphonuclear leukocytes and resists complement-mediated bactericidal activity,

153. The answer is a. *(McPhee, 2/e, p 61.)* Neisseria meningitidis is a common cause of meningitis in children older than 2 months of age. Among older children, adolescents, and adults, *N. meningitidis* and *S. pneumoniae*

are the two most common pathogens of meningitis and, among older children and adolescents, *H. influenzae* is also common. *Escherichia coli* (and other gram-negative bacilli), group B streptococci and other streptococci, and *Legionella monocytogenes* commonly cause meningitis in children less than 2 months of age.

154. The answer is b. (*McPhee, 2/e, p 64.*) *Streptococcus pneumoniae* is the most common cause of community-acquired pneumonia, accounting for about two-thirds of pneumonias, especially among adults. *Mycoplasma pneumoniae* occurred mainly in young adults during the second and third decades. *Haemophilus influenzae* is a frequent cause of community-acquired pneumonia, but not the most frequent. *Staphylococcus aureus* and *Legionella* spp. are minor causes of community-acquired pneumonia.

155. The answer is a. (*McPhee, 2/e, p 65.*) *Mycoplasma pneumoniae* shows no more predilection for HIV and AIDS patients than it does for persons with intact immune systems. *Pneumocystis carinii* pneumonia (PCP) occurs almost exclusively in persons with HIV and AIDS and not in persons with intact immune systems; it usually occurs when the CD4 lymphocyte count falls below about 450 cells/µL. However, prophylaxis with trimethoprim–sulfamethoxazole or pentamidine aerosols now prevent most of the cases of PCP. *Mycobacterium tuberculosis* occurs at a high rate in persons with AIDS; they can become infected and spread drug-resistant strains of *M. tuberculosis* that are very difficult to treat especially if they exhibit resistance to both isoniazid and rifampin. *Streptococcus pneumoniae* and *H. influenzae* occur at high rates in persons with HIV and AIDS.

156. The answer is e. (*McPhee, 2/e, p 65.*) *Mycobacterium tuberculosis* is an airborne microorganism that is deposited directly into the lower airways. The large microorganisms are caught in the nose and pharynx and colonize these structures, and the smaller microorganisms are deposited on the mucociliary blanket of the respiratory tree.

157. The answer is b. (*McPhee, 2/e, p 65.*) *Pneumocystis carinii*, a common cause of pneumonia in persons with HIV and AIDS, was the most common cause of pneumonia in this group of persons until prophylaxis either with trimethoprim–sulfamethoxazole or pentamidine aerosols became a routine feature of their care.

Infectious Disease   Answers   57

158. The answer is d. *(McPhee, 2/e, p 65.)* Persons with acute and chronic alcoholism become infected with *Klebsiella pneumoniae*, most often by aspiration, and *K. pneumoniae* pneumonia often localizes in the upper lobe, sometimes the minor fissures bows downward, reflecting the bogginess of the upper lobe involved with this infection.

159. The answer is c. *(McPhee, 2/e, p 65.)* Psittacine birds transmit *Chlamydia psittaci* to humans who come in contact with them or keep them as pets. The pneumonia appears on chest roentgenogram as other atypical pneumonias, such as *M. pneumoniae* pneumonia. It also responds to treatment with tetracycline.

160. The answer is e. *(McPhee, 2/e, p 65.)* Persons suffering from chronic lung disease develop pneumonia due to *Moraxella catarrhalis*, and also to *Streptococcus pneumoniae* and *H. influenzae*. Very often, their sputum is colonized by one or more of the microorganisms that gain entrance to the lower respiratory tract because of damage to the respiratory tract cilia and the mucociliary blanket.

161. The answer is a. *(McPhee, 2/e, p 65.)* Transplant recipients are at risk for pneumonia from a number of different microorganisms, including *Legionella* spp., as well as cytomegalovirus, *Aspergillus*, and occasionally *P. carinii*.

162. The answer is d. *(McPhee, 2/e, p 67.)* Viruses account for 30 to 40 percent of the cases of infectious diarrhea in the United States; rotavirus is the predominant virus, especially in infants and children. The main bacterial pathogens are *Helicobacter pylori* and various *E. coli* serotypes. *Cryptosporidium* causes a particularly severe diarrhea in persons with HIV and AIDS and a very low CD4 lymphocyte count, usually less than 100 cells/µL.

163. The answer is d. *(McPhee, 2/e, p 68.)* On a worldwide basis, the single most important bacteria that causes diarrhea is *E. coli*, of which several main types play important roles in diarrhea illnesses. These include the following types: enteroaggregative, enteropathogenic, enterotoxigenic, enteroinvasive, and enterohemorrhagic. Enterotoxigenic bacteria occur more widely in acute diarrhea than in the other types. They produce two enterotoxins that adversely affect the mucosal cells of the small intestine, resulting in a watery diarrhea. Enterohemorrhagic *E. coli* of serotype O157:H7 is

58   Pathophysiology

associated with the severe hemolytic–uremia syndrome, as well as non-bloody diarrhea, bloody noninflammatory diarrhea, and thrombotic thrombocytopenia purpura. It also produces Shiga-like cytotoxins that are comprised of one large protein unit (the A subunit) and five small subunits (the B subunit) that bind the toxin to the intestinal cell, stop intracellular protein synthesis, and eventually kill the intestinal cells.

164. The answer is a. *(McPhee, 2/e, p 68.)* Certain bacteria can aggregate at the mucosal wall as a mechanism of producing diarrhea, but an antigen–antibody reaction is involved. Enterotoxin, cytotoxin, and invasion of the mucosal wall are mechanisms of bacterial diarrhea in differing bacterial pathogens.

165. The answer is e. *(McPhee, 2/e, p 75.)* Exotoxin is not an endogenous mediator of sepsis. The other chemicals are mediators of sepsis; additional mediators include platelet-activating factor, endothelium-derived relaxing factor, kinin, coagulation, and myocardial depressant substance.

166. The answer is d. *(McPhee, 2/e, p 44.)* CD4 T lymphocyte is the main CD antigen involved in HIV and AIDS. HIV infection destroys the CD4 lymphocytes. The CD8 lymphocytes increase in a reciprocal manner in HIV.

167. The answer is e. *(McPhee, 2/e, p 44.)* Bone marrow stimulation is not involved; viral proteins show toxicity not only for the CD4 lymphocytes, but also for the marrow, suppressing its function. Apoptosis (programmed cell death), autoimmune destruction of CD4 lymphocytes, and syncytium formation contribute to the decrease in CD4 lymphocytes.

168. The answer is a. *(McPhee, 2/e, p 44.)* Eosinophils do not play a role in the pathogenesis of HIV and AIDS. The other cells do. Monocytes and macrophages transfer virus to sites of infection and monocytes also release cytokines.

169. The answer is d. *(McPhee, 2/e, p 44.)* Usually, HIV progresses slowly, and the beginning of immunosuppression is about 5 to 10 years after the onset of infection. Infection begins with an acute, brief, febrile viral syndrome, followed by a long symptom-free period until the CD4 T lymphocyte begins to decline, providing evidence of immunosuppression.

Infectious Disease   Answers   59

170–177. The answers are g, a, e, d, j, k, b, h. (*McPhee, 2/e, pp 45–48.*) *Pneumocystis carinii* pneumonia is a common complication of HIV and AIDS, and it is the most common opportunistic infection in HIV and AIDS. It can be prevented in most persons by prophylaxis with trimethoprim–sulfamethoxazole or pentamidine aerosols. A serious lung infection, it can be life threatening. *Candida albicans* causes thrush of the mouth and a severe esophagitis that causes pain and dysphagia. Biliary tract disease of differing types can be caused by *Cryptosporidium*, a protozoa that infects the gastrointestinal tract in many patients with HIV and AIDS. Herpesvirus infections are common in persons with HIV and AIDS. Many of them experience reactivation of latent herpesvirus varicella infections (shingles) that are painful and potentially can communicate chickenpox to susceptible persons. Cytomegalovirus, also a herpesvirus, causes retinitis that, without treatment, leads progressively to complete blindness. Toxoplasmosis, acquired from infected cats, causes space-occupying lesions, accompanied by central nervous system symptoms. Wasting and cachexia are caused by a bloodstream infection with *Mycobacterium avium–intracellulare* complex. Kaposi's sarcoma typically causes purplish, dense skin lesions of varying size and also visceral organ involvement. All persons with HIV and AIDS experience one or more of these complications, especially as the immune system progressively fails.

178. The answer is c. (*Fauci, 14/e, p 1045.*) Rickettsiae are gram-negative coccobacilli transmitted by insect vectors. However, not all are transmitted by insect bites. Epidemic and endemic typhus are transmitted through feces scratched into the bite lesion or scratch mark; Q fever and trench fever are transmitted through inhalation.

179. The answer is e. (*Fauci, 14/e, p 1036.*) All of these are manifestations of leptospirosis. Onset of severe disease (Weil's syndrome) occurs in a similar manner to mild disease but progresses to renal and vascular dysfunction.

180. The answer is c. (*Fauci, 14/e, pp 11 and 30–31.*)

181. The answer is e. (*Fauci, 14/e, p 1178.*) Bacterial diarrheas can all mimic intestinal amebiasis, including enteroinvasive *E. coli* and *Vibrio*. Usually, the parasitic manifestation causes less fever but does produce few neutrophils and heme-positive stools.

60 Pathophysiology

182. The answer is d. *(Fauci, 14/e, p 1025.)* Primary syphilis is associated with regional lymphadenopathy such as inguinal nodes and with a vaginal or penile chancre. Generalized lymphadenopathy is associated with secondary syphilis.

183. The answer is e. *(Fauci, 14/e, p 1013.)* Prophylaxis is recommended for low-risk groups less than 35 years of age with PPD ≥ 15 mm, for high-risk groups with PPD ≥ 10 mm (less than 35 years of age), and also for persons with a high-risk medical condition and PPD ≥ 10 mm.

184. The answers are f, f, t, t, t.

185. The answers are f, f, t. *(Fauci, 14/e, p 1093.)* A fourfold or greater increase in antibody titer along with the isolation of the virus is a reliable diagnostic approach also. Excretion in the urine can continue chronically.

186. The answer is f. *(Fauci, 14/e, p 1186.)* Doxycycline and tetracycline should not be used alone but may be used in combination. It is important to determine areas of drug resistance, so medication should be chosen accordingly. Quinine and quinidine should be given if resistance is suspected.

187. The answers are f, f, t, t, t. *(Fauci, 14/e, p 985.)* Most cases occur in children and do not cause fever in the majority of cases. This is a mostly self-limiting disease, and efficacy of antibodies is unclear.

188. The answers are a-t, b-f, c-t, d-t, e-t. *(Fauci, 14/e, p 878.)* TSS is a life-threatening illness and must be differentiated from other illnesses causing hypotension, rash, and fever, such as Rocky Mountain spotted fever, severe drug reactions, Kawasaki syndrome, and gram-negative sepsis. Fifty percent of cases occur in nonmenstrual settings.

Cardiovascular

Questions

DIRECTIONS: Each item below contains a question or incomplete statement followed by suggested responses. Select the **one best** response or the **matching** response to each question.

189. Which one of the following ECG components varies with heart rate?
a. PR interval
b. QRS duration
c. ST segment
d. QT interval
e. None of these

190. Which of the following cardiac parameters increases during pregnancy?
a. Cardiac output
b. Stroke volume
c. Heart rate
d. Blood volume
e. All of these

191. Which of the following does NOT occur during diastole?
a. Blood passes from atria into the ventricles.
b. The atrioventricular (AV) valves are open.
c. Rapid ventricular filling occurs.
d. The ventricles contract.
e. Atrial contraction propels final proportion of blood into ventricles.

192. Which of the following is NOT a component of cardiac output?
a. Pulse pressure
b. Heart rate
c. Myocardial contractility
d. Preload
e. Afterload

Pathophysiology

Items 193–195

For the Frank-Starling curve, match the appropriate curve with the following description:

193. A normal individual whose stroke volume increases as preload increases

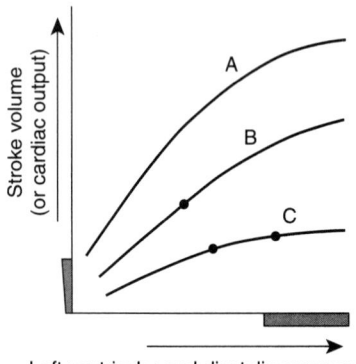

194. A patient in congestive heart failure due to systolic dysfunction

195. A patient with normal left ventricular function who is receiving intravenous (IV) dobutamine as part of a diagnostic study for ischemia

Items 196–199

Identify the points on the normal left ventricular pressure volume loop:

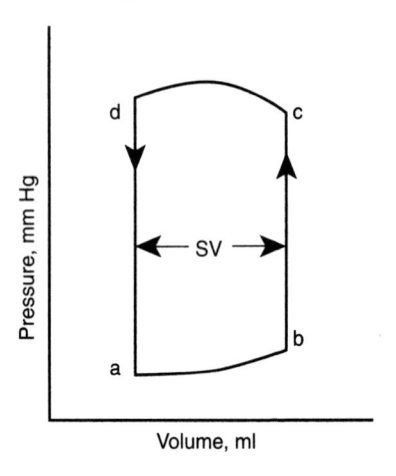

196. Mitral valve opens

197. Mitral valve closes

198. Aortic valve opens

199. Aortic valve closes

Items 200–203

Match the appropriate complex or waveform on the ECG tracing with the following description:

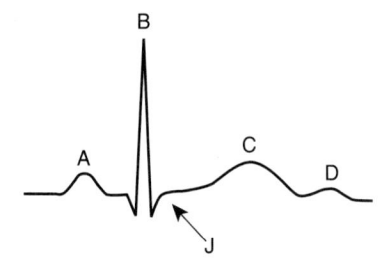

[ECG tracing labeled A, B, C, D, with J pointing to a notch after B]

200. Would reflect left or right atrial enlargement

201. Often present during hypokalemia

202. Represents repolarization of ventricles

203. Would widen if a bundle branch block were present

204. A normal frontal plane QRS axis is
 a. +90° to +180°
 b. −30° to −90°
 c. −30° to +90°
 d. 0° to +150°
 e. 0° to +90°

205. Which of the following associations is correct?
 a. Hypercalcemia: shortened QT interval
 b. Hypocalcemia: prolonged QT interval
 c. Hyperkalemia: peaked T waves
 d. Hypokalemia: U waves
 e. All of these

Items 206–210

Match the appropriate electrolyte flux with the correct phase of the following action potential in a normal ventricle:
a. Influx of sodium ions
b. May involve chloride ion movement
c. Rapid potassium exit
d. Resting state
e. Mediated via slow calcium channels

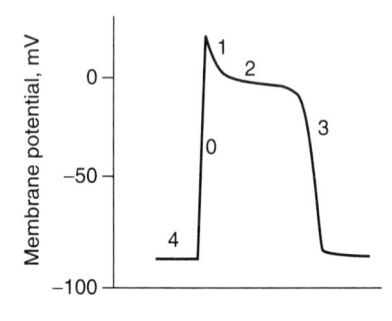

206. Phase 0

207. Phase 1

208. Phase 2

209. Phase 3

210. Phase 4

211. Which of the following arterial pulse waveforms is consistent with severe left ventricular impairment?
a. Parvus et tardus pulse
b. Bisferiens pulse
c. Pulsus alternans
d. Hyperkinetic pulse
e. Dicrotic pulse

212. Which of the following arteriole pulse waveforms is consistent with aortic stenosis?
a. Pulsus alternans
b. Pulsus tardus
c. Bisferiens pulse
d. Dicrotic pulse
e. Parvus et tardus pulse

213. Which of the following conditions is consistent with a hypokinetic arterial pulse?
a. Left ventricular failure
b. Hypovolemia
c. Restrictive pericardial disease
d. Mitral stenosis
e. All of these

214. Which of the following statements is true of a reversed splitting of the first heart sound?
a. The mitral component follows the tricuspid component.
b. It may be present in severe mitral stenosis.
c. It may be present with a left atrial myxoma.
d. It may be present with a left bundle branch block.
e. All of these.

Items 215-220

Match the following heart sounds with the appropriate description:
a. low-pitched sound produced in the ventricle at the termination of rapid filling, heard in normal children and in patients with increased cardiac output
b. high-pitched, early diastolic sound, usually due to mitral stenosis
c. produced by closure of the AV valves
d. produced by closure of the semilunar (aortic and pulmonic) valves
e. low-pitched, presystolic sound of ventricular filling produced by atrial contraction
f. often caused by mitral or tricuspid valve prolapse

215. S_1

216. S_2

217. S_3

218. S_4

219. Opening snap

220. Midsystolic click

221. The onset of the QRS complex on surface ECG corresponds to which action potential phase?
a. Phase I
b. Phase II
c. Phase III
d. Phase IV
e. Phase 0

222. The isoelectric ST segment on surface ECG corresponds to which action potential phase?
a. Phase I
b. Phase II
c. Phase III
d. Phase IV
e. Phase 0

223. The T wave on the surface ECG corresponds to which action potential phase?
a. Phase I
b. Phase II
c. Phase III
d. Phase IV
e. Phase 0

224. Which is present when no atrial impulse conducts to the ventricles?
a. First-degree AV block
b. Second-degree AV block type I
c. Second-degree AV block type II
d. Third-degree AV block
e. None of these

225. Which of the following AV blocks is characterized by progressive PR interval prolongation prior to loss of AV conduction?
a. First-degree AV block
b. Second-degree AV block type I
c. Second-degree AV block type II
d. Third-degree AV block
e. None of these

226. Which of the following is NOT true of left ventricular ejection fraction?
a. It is defined as the fraction of end-diastolic volume ejected per beat.
b. It equals stroke volume divided by end-diastolic volume.
c. It is impaired in a patient suffering from congestive heart failure due to systolic dysfunction.
d. Its normal range is 55 to 75 percent.
e. None of these.

227. Which of the following is the correct sequence for myocardial depolarization?
a. AV node → bundle of His → atria
b. Bundle of His → AV node → left ventricle →
c. Sinoatrial (SA node) → AV node → bundle of His → right and left ventricles
d. SA node → left ventricle → bundle of His
e. None of these

228. Which ECG leads represent the inferior cardiac wall?
a. V_1, V_2
b. V_3, V_4
c. aVR
d. I, aVL
e. II, III, aVF

229. Which of the following produces a diastolic murmur?
a. Aortic regurgitation
b. Aortic stenosis
c. Mitral regurgitation
d. Supravalvular aortic stenosis
e. Tricuspid regurgitation

230. Loss of P waves on surface ECG is consistent with
a. first-degree AV block
b. atrial flutter
c. atrial fibrillation
d. sinus bradycardia
e. second-degree AV block type I

231. The arrow indicates

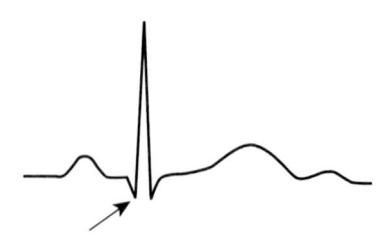

a. R wave
b. S wave
c. QS wave
d. Q wave
e. T wave

232. Cardiac output is the product of
a. preload × stroke volume
b. afterload × heart rate
c. heart rate × stroke volume
d. contractility × preload
e. None of these

233. Which of the following is true regarding right ventricular hypertrophy?
a. The hypertrophy may result from pulmonic valve stenosis.
b. The hypertrophy is characterized by a tall R wave and leave V_1 on the surface ECG.
c. The hypertrophy is usually associated with right axis deviation.
d. The hypertrophy may be secondary to an atrial septal defect.
e. All of these.

Cardiovascular

Answers

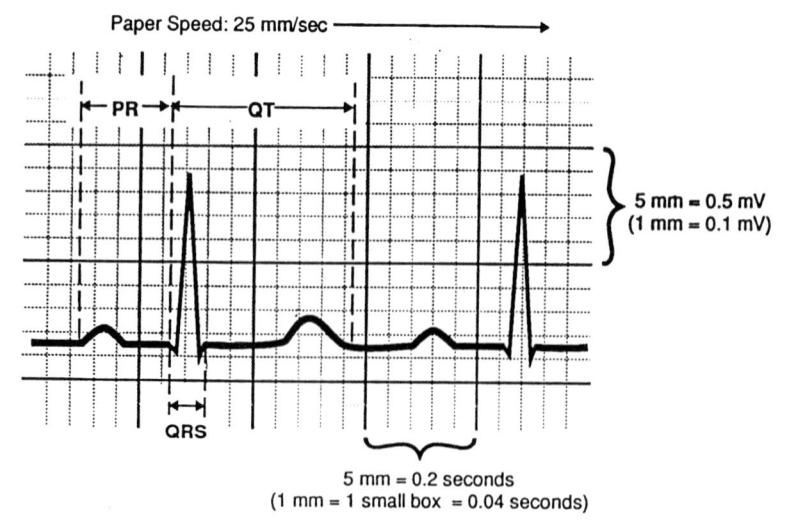

189. The answer is d. *(Fauci, 14/e, p 1238.)* The standard surface ECG is divided into various intervals and segments. The small horizontal "boxes" on a standard ECG each equal 0.04 s, five small "boxes" comprise one large box, and thus the large "box" measures 0.20 s. The *PR interval* is measured from the beginning of the P wave to the beginning of the QRS complex and represents AV conduction. A normal PR interval measurement is 0.12 to 0.20 s. A PR interval greater than 0.20 s is referred to as a first-degree AV block. The *QRS complex* on a standard surface ECG represents ventricular depolarization. Normal QRS duration is <0.10 s. If the QRS duration is >0.10 s, a bundle branch block is most likely present. The *ST segment* is the most usual site evaluated for the presence of ischemia or injury. The ST segment is elevated in acute myocardial injury and depressed in the presence of myocardial ischemia. The *QT interval* varies with heart rate. This interval represents both ventricular depolarization and repolarization. The QT interval increases with bradycardia and decreases as the heart rate increases. The normal QT interval depends on heart rate, but a corrected "QT interval" may be calculated as the QT interval divided by the square root of the R–R interval. The normal corrected QT interval is <0.44 s. Prolongation of the QT interval is clinically important, because it can lead to fatal dysrhythmias such as torsades de pointes.

190. The answer is e. *(Fauci, 14/e, p 26.)* Normal cardiovascular changes that occur with pregnancy include decreased systemic vascular resistance, increased blood volume, increased stroke volume, increased heart rate, and increased cardiac output. These changes are quite well tolerated during normal pregnancy but, with preexisting cardiac disease, these changes may not be well tolerated. Because of these normal cardiovascular changes during pregnancy, new systolic murmurs may develop, as well as the presence of a third heart sound. In a physiologically normal heart, these developments are considered normal during pregnancy.

191. The answer is d. *(Lilly, pp 18–19.)* The cardiac cycle consists of both systole and diastole. During diastole, the AV valves are open and the atrial and ventricular pressures are essentially equal. Left atrial contraction occurs in late diastole, causing a small increase in pressure in the left atrium and left ventricle. Left ventricular systole initiates left ventricular contraction. The mitral valve closes as left ventricular pressure overcomes left atrial pressure; this represents the mitral component of the first heart sound. As left ventricular pressure continues to increase, the aortic valve opens. As the ventricles then begin to relax, the pulmonic and aortic valves close, causing the second heart sound. As left ventricular pressure continues to drop, the aortic valve opens. The contraction of both ventricles initiates systole and does not occur during diastole.

192. The answer is a. *(Lilly, pp 148–150.)* Cardiac output, which is the volume of blood ejected from the ventricle in 1 min, changes regularly as the body's needs change and is the product of stroke volume and heart rate. Therefore, for instance, when an individual begins to exercise and heart rate increases, the cardiac output will increase as well. Stroke volume is merely the volume of blood ejected from the ventricle during systole. Preload is the ventricular wall tension present at the end of diastole. Afterload is the ventricular wall tension present during contraction. Contractility is actual

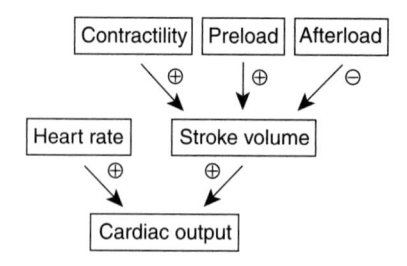

70   Pathophysiology

```
    ┌─────────────┐ ┌─────────┐ ┌──────────┐
    │Contractility│ │ Preload │ │ Afterload│
    └─────────────┘ └─────────┘ └──────────┘
              ⊕         ⊕          ⊖
    ┌───────────┐      ┌───────────────┐
    │ Heart rate│      │ Stroke volume │
    └───────────┘      └───────────────┘
              ⊕         ⊕
              ┌───────────────┐
              │ Cardiac output│
              └───────────────┘
```

measurement of the strength of contractile force. Contractility and preload contribute positively to stroke volume, whereas increased afterload decreases stroke volume. Thus, ultimately contractility, preload, afterload, stroke volume, and heart rate are all contributors to cardiac output.

193–195. The answers are b, c, a. *(Lilly, p 150.)* Frank-Starling curves or ventricular function curves are diagrams that show the relationship between stroke volume or cardiac output and preload or left ventricular end-diastolic volume. In normal individuals, as left ventricular end-diastolic pressure or preload increases, stroke volume will increase proportionately. In patients who suffer heart failure, increased left ventricular end-diastolic pressure is not met with increased stroke volume, because the contractility is depressed and is unable to function; thus, the patient ultimately experiences heart failure. If an individual with normal contractility is administered a positive inotropic agent, such as dobutamine, the contractility will increase without a necessary increase in preload. This is reflected by curve (A) on the Frank-Starling curve. The middle curve (B), represents a normal

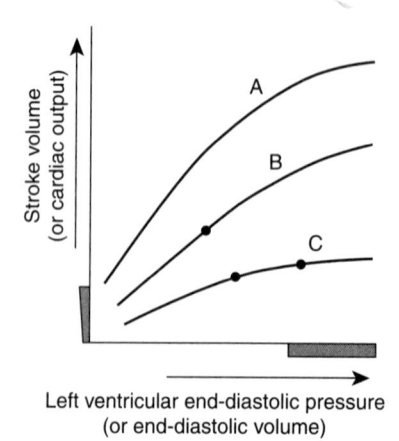

```
  Stroke volume
  (or cardiac output)
       ↑
       │        _____  A
       │       /
       │      /     _____  B
       │     /    /
       │    •   /   _____  C
       │       •   /
       │         •
       └──────────────────→
       Left ventricular end-diastolic pressure
              (or end-diastolic volume)
```

individual whose stroke volume will increase as preload increases. The bottom curve (C) represents a patient with congestive heart failure in that increasing this individual's preload will not be met with an increase in stroke volume and will result in pulmonary edema.

196–199. The answers are a, b, c, d. (*Lilly, p 151.*) A normal left ventricular volume loop is represented here. The mitral valve opens at point A. Diastolic filling ensues, represented by the line A–B, at which point ventricular contraction begins and the mitral valve closes at point B. Isometric contraction (the aortic valve is not yet open) is represented by the line B–C indicating the increase in pressure in the left ventricle. As the left ventricular pressure maximizes, the aortic valve opens at point C. Line C–D thus corresponds with left ventricular ejection, and the aortic valve closes at point D. Line D–A thus represents isometric relaxation and subsequently reopening of the mitral valve at point A.

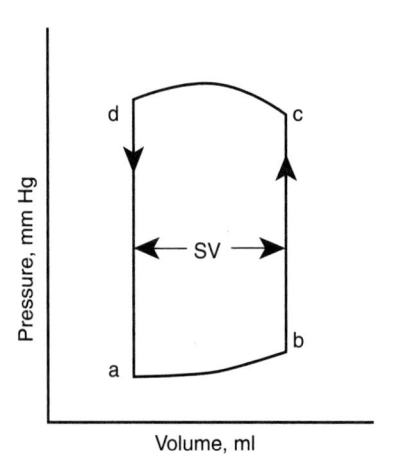

200–203. The answers are a, d, c, b. (*Fauci, 14/e, pp 1238–1242.*) The waveforms on a standard surface ECG include the P wave, which precedes the QRS complex. It is usually a rather small upward deflection that represents atrial depolarization. Thus, if right or left atrial enlargement were present, the P wave would reflect these changes. During right atrial enlargement, the P wave is tall, as seen in lead II. During left atrial enlargement, the P wave tends to become widened and biphasic, which is best seen in leads V_1 and V_2. Electrolyte abnormalities are common causes of changes on the surface of an

72   Pathophysiology

ECG. Hypokalemia results in the presence of U waves, which is a relatively low devoltage deflection after the T wave. U waves are nonspecific and may be present for other reasons, but they are a classic finding in hypokalemia. The QRS complex is usually the largest complex on the surface tracing. This complex represents depolarization of the left and right ventricles. Normal QRS duration is <0.1 s. If a right or left His bundle is blocked, this results in a widened QRS complex representing abnormal ventricular depolarization. The third wave form on a standard surface ECG is the T wave, which represents repolarization of the ventricles. The T wave may show changes during myocardial ischemia, electrolyte abnormalities, and may be altered by many pharmacologic agents.

204. The answer is c. *(Lilly, pp 57–67.)* The standard surface ECG is recorded from 12 leads. Leads V_1 through V_6 are placed on the anterior chest and are referred to as the chest leads. Leads aVR, aVF, and aVL are unipolar limb leads and are averaged together to create a standard reference. The aVR selects the right arm as the positive electrode, the aVF selects the left leg as the positive electrode, and the aVL selects the left arm as the positive electrode. Leads I, II, and III are also limb leads but are bipolar. Lead I designates the left arm as the positive electrode and the right arm as negative. Lead II has the left leg designated as the positive electrode and the right arm as the negative electrode. Lead III has the left arm designated as the negative electrode and the left leg as the positive electrode. When the six limb leads are interposed, a reference system is devised. The main QRS electrical axis may be determined by averaging the forces created during ventricular depolarization. The normal frontal plane QRS axis is −30 to approximately +90. An axis more negative than −30° indicates left axis deviation, and an axis greater than +90° indicates right axis deviation. The axis is determined by the positive or negative direction of the QRS complex in each of the limb leads.

205. The answer is e. *(Lilly, p 79.)* Electrolyte abnormalities affect various portions of the surface ECG. Both hypercalcemia and hypocalcemia affect ventricular repolarization and are thus represented by changes in the QT interval. Hypercalcemia results in a shortened QT interval, whereas hypocalcemia results in a prolonged QT interval. Calcium does not affect the T wave; it specifically changes the ST portion of the QT interval. Hyperkalemia may be represented by very tall peaked T waves as potassium affects ventricular repolarization. Hypokalemia may be represented by U waves, which are small deflections following the T wave.

206–210. The answers are a, b, e, c, d. (*Lilly, p 12.*) Myocardial contraction ultimately results from electrical impulses. An action potential is created by ion fluxes through certain channels in the cellular membranes. Various cardiac cells are capable of electrical activity including the pacemaker cells of the SA and AV nodes, the Purkinje fibers, and the cardiac muscle cells. The action potential consists of phases I to IV and of phase 0. Phase 0 is caused by the rapid influx of sodium ions. This is the rapid depolarization phase of the action potential. Phase I is not well understood but is the first stage of repolarization. This phase is thought to include chloride ion movement. Phase II is controlled by the slow calcium channels and is commonly referred to as the plateau. The slow calcium influx is an important factor in myocyte contraction. Phase III consists largely of the exiting of potassium and returning of the resting potential to approximately –90 mV. Phase IV is simply the resting state prior to stimulation.

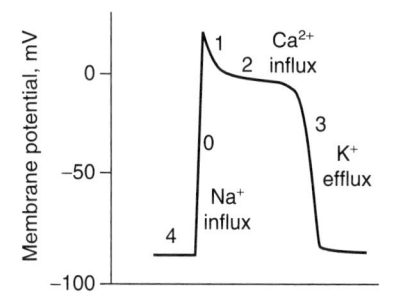

211–213. The answers are c, b, e. (*Fauci, 14/e, p 1232.*) The arterial pressure pulse may be palpitated in the periphery, or evaluation of the carotid pulse may occur. Certain changes in the arterial pressure pulse occur with various pathologic conditions. Pulsus parvus, defined as a small weak pulse is present when left ventricular stroke volume is decreased. A hypokinetic or weak pulse is commonly present due to conditions such as hypervolemia, heart failure, restrictive pericardial disease, or mitral stenosis. Pulsus tardus (late pulse), during which the systolic peak is delayed. Is common in aortic stenosis, because the left ventricular ejection is impeded due to the stenotic valve. A bisferiens pulse has two peaks and is common with aortic regurgitation. Pulsus alternans is a unique pattern during which the amplitude of the pulse changes or alternates in size with a stable heart rhythm. This is common in severe left ventricular dysfunction. In summary, examination of the arterial pressure pulse may lead to clues of existing pathology.

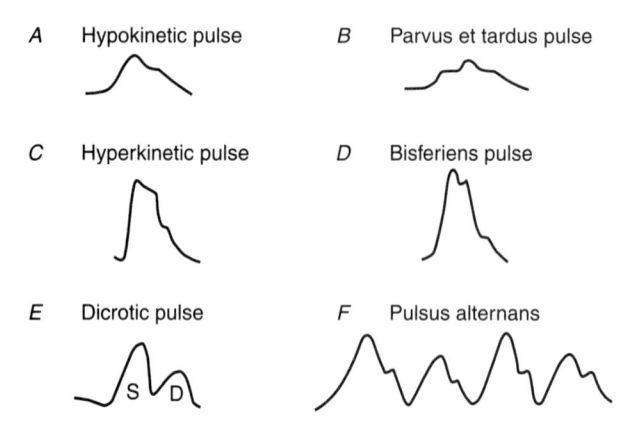

A Hypokinetic pulse   B Parvus et tardus pulse

C Hyperkinetic pulse   D Bisferiens pulse

E Dicrotic pulse   F Pulsus alternans

214. The answer is e. (*Fauci, 14/e, p 1234.*) The first heart sound consists of a mitral and tricuspid component. Normally, the mitral valve closes first, thus contributing to the first component of S_1. When there is reverse splitting of S_1, the mitral component follows the tricuspid component. This finding may be present in severe mitral stenosis and may be due to the presence of a left atrial myxoma, which often mimics mitral stenosis and may be present with a left bundle branch block.

215–220. The answers are c, d, a, e, b, f. (*Fauci, 14/e, pp 1234–1235.*) The *first heart sound* is produced by closure of the mitral and tricuspid valves. Normally, the mitral valve closure precedes the tricuspid valve closure; thus, the mitral valve contributes to the first component of S_1. The *second heart sound* is caused by the aortic and pulmonic valve closure. The aortic component is generally louder than the pulmonic component. The fixed splitting of the second heart sound is caused by atrial septal defect. Normally, the splitting of S_2 representing the difference in closure between the aortic and pulmonic valves varies with respiration. If an atrial septal defect is present, this variation does not occur and the two components of S_2 are "fixed." A *third heart sound* is present in individuals with increased cardiac output. It is a low-pitched sound produced in the ventricle. In adults, this is a pathologic

Cardiovascular   Answers   75

finding, but an S_3 is quite normal in young children. A *fourth heart sound* may be present and is produced by atrial contraction. The S_4 is a low-pitched sound produced during ventricular filling. It may be a normal finding in the elderly, because the left ventricle tends to "stiffen" with age. Obviously, because the sound is due to atrial contraction, it is absent in patients with atrial fibrillation. A classic finding of mitral stenosis is an *opening snap* that is a high-pitched early diastolic sound. An opening snap may be noted with tricuspid stenosis. *Midsystolic clicks* usually result from mitral or tricuspid prolapse and are due to unequal chordae tendinae.

221–223. The answers are e, b, c. *(Fauci, 14/e, p 1238.)* The surface ECG tracing, which is a representation of the electrical activity of the heart, corresponds to the various phases of the ventricular action potential. The QRS complex on surface ECG represents ventricular depolarization. The intracellular activity during ventricular repolarization is a rapid influx of sodium and corresponds to phase 0 of the action potential. The isoelectric ST segment corresponds to continued ventricular depolarization and repolarization. This corresponds to the plateau phase or phase II. This phase is mediated via slow calcium channels. The T wave on surface ECG represents ventricular repolarization. Intracellularly, this corresponds to phase III, during which potassium rapidly exits the cells.

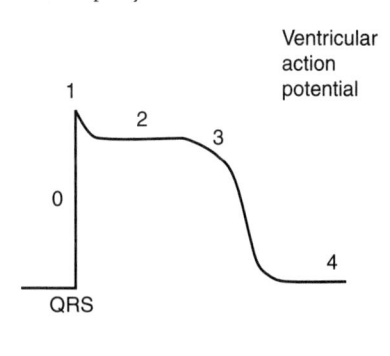

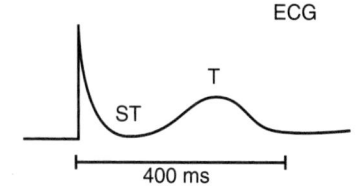

76　Pathophysiology

224. The answer is d. *(Fauci, 14/e, p 1256.)* *First-degree AV block* is represented by a prolonged PR interval. This is due to delayed AV conduction. *Second-degree AV* block is divided into two types: I and II. Type I is an AV block above the level of the His bundle and is characterized by gradually prolonging PR intervals followed by a P wave that is not conducted to the ventricle. Second-degree AV block type II is generally due to disease of the His–Purkinje system. The PR interval does not gradually prolong, and there is sudden loss of a QRS complex following a P wave. *Third-degree block* is present when no AV conduction occurs. The atrium is depolarizing independently of the ventricle.

225. The answer is b. *(Fauci, 14/e, p 1256.)* Please see the preceding explanation.

226. The answer is e. *(Lilly, p 150.)* Left ventricular ejection fraction is defined as the fraction of end-diastolic volume ejected from the ventricle per beat. Normal adult ejection fraction is 55 to 75 percent. In patients with impaired contractility or congestive heart failure due to systolic heart dysfunction, the ejection fraction may be markedly impaired. Left ventricular ejection fraction is calculated by dividing stroke volume by end-diastolic volume.

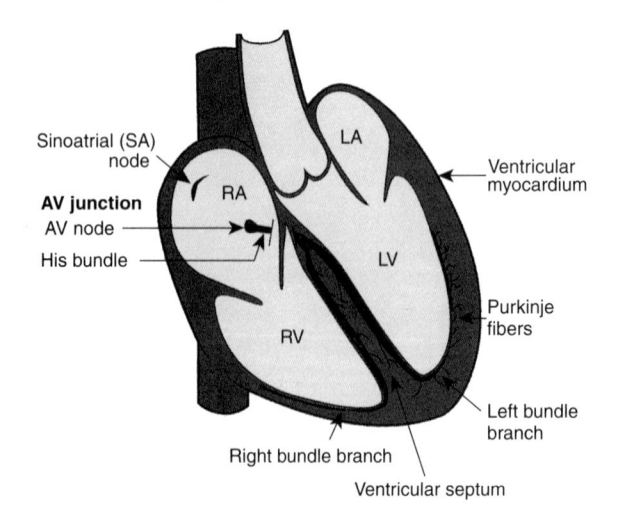

227. The answer is c. *(Fauci, 14/e, pp 1237–1238.)* The SA node, which is the site of initiation of a depolarization that results in a normal heartbeat, consists of pacemaker cells that fire spontaneously. The impulse then proceeds to the conduction tissues in the AV node in His bundle, both of which are in the AV junction. The bundle of His then divides into the right bundle and left bundle, and the impulse is then conducted to the right and left ventricular myocardium through the Purkinje fibers. The left bundle bifurcates into left anterior fascicle and left posterior fascicle. The depolarization then continues through both ventricular walls, and ventricular contraction is triggered. Depolarization occurs from endocardium to epicardium. Thus, the correct sequence for myocardial depolarization is SA node through AV node through bundle of His and proceeding to both ventricles.

228. The answer is e. *(Lilly, p 76.)* The lead system on a standard surface ECG is designed so that certain leads represent specific segments of the left ventricle. The chest (or precordial) leads represent the anterior wall of the left ventricle with leads V_3 and V_4, the septal portion of the left ventricle with leads V_1 and V_2, and the low lateral portion of the left ventricle with leads V_5 and V_6. The high lateral wall is represented by the limb leads I and aVL. The inferior wall is represented by II, III, and aVF. Lead aVR does not represent a specific portion of the left ventricular muscle, because its positive electrode is the right arm.

229. The answer is a. *(Fauci, 14/e, p 1236.)* Systole is represented by left ventricular contraction and rapid ejection. Left ventricular systole must occur when the aortic valve is open. If the aortic valve is stenosed significantly, a systolic murmur will ensue as the blood flows through the constricted valve. If the mitral valve is incompetent or weakened during systole or left ventricular ejection, there will be backflow into the left atrium through the incompetent valve, and *mitral regurgitation* will result. As the backflow continues into the left atrium during systole, mitral regurgitation will result in a *systolic murmur*. Supravalvular aortic stenosis occurs when there is obstruction above the aortic valve structure proper and, again, the flow constriction will occur during ventricular systole, thus resulting in a systolic murmur. Tricuspid regurgitation, similar to mitral regurgitation, will also result in a systolic murmur due to backflow into the right atrium during right ventricular systole. In contrast, *aortic regurgitation* results in backflow into the left ventricle during relaxation. If the aortic valve is incompetent, as opposed to being stenosed, as the ventricle relaxes and the aortic valve is closed there will be regurgitation resulting in a *diastolic murmur*.

78 Pathophysiology

230. The answer is c. *(Lilly, p 71.)* Atrial depolarization is represented on the surface ECG as a P wave. Left or right atrial enlargement that exists is represented with increased voltage of the P wave. Atrial fibrillation occurs when there is chaotic, intermittently conducted atrial activity. Because there is no organized atrial contraction, no P wave is represented on the surface ECG.

231. The answer is d. *(Fauci, 14/e, p 1238.)* The QRS complex on a surface ECG represents ventricular depolarization. If the initial force of this complex is in a negative direction, this is referred to as a Q *wave*. The subsequent upward deflection is an *R wave* and the final downward deflection is an *S wave*. A Q wave need not necessarily be present, if the initial force of the QRS complex is positive, then a Q wave by definition is not present. Pathologic Q waves occur when a myocardial infarction has occurred. Small Q waves are usually nonpathologic.

232. The answer is d. *(Lilly, pp 148–149.)* Cardiac output, which is defined as the volume of blood expelled from the ventricle per minute, is the product of heart rate and stroke volume. In turn, stroke volume or the volume (blood that the ventricle ejects in systole) is determined by contractility, preload, and afterload.

233. The answer is e. *(Fauci, 14/e, p 1241.)* A cardiac hypertrophy of the various chambers is manifested on the surface ECG. If pulmonic valve stenosis is present, right ventricular pressure is significantly increased as the ventricle must contract against a stenotic valve. This results in hypertrophy of the right ventricle. The classic surface ECG findings include a prominent R wave in lead V_1 or V_3, with a large R wave associated. Right ventricular hypertrophy also often leads to ST depression and T-wave changes in the precordial leads. This is often referred to as secondary repolarization changes or "strain pattern." If an atrial septal defect is present, the right ventricle is significantly volume overloaded, again resulting in right ventricular hypertrophy.

PULMONARY

Questions

DIRECTIONS: Each item below contains a question or incomplete statement followed by suggested responses. Select the **one best** response to each question.

234. All of the following are features of primary pulmonary hypertension (PPH) EXCEPT
a. It is more common in females.
b. The findings of pulmonary function tests are usually normal.
c. Clubbing is not a feature of PPH.
d. Chronic obstructive pulmonary disease (COPD) is a cause of PPH.
e. Treatment includes vasodilators and anticoagulation.

235. A distinction between transudative and exudative effusion is based on
a. a pleural fluid protein/serum protein ratio of less than 0.5
b. a pleural fluid LDH/serum LDH ratio of less than 0.6
c. a pleural fluid LDH more than two-thirds of the normal upper limit for serum
d. all of these
e. none of these

236. The following are common causes of transudative pleural effusion EXCEPT
a. congestive heart failure (CHF)
b. cirrhosis
c. malignant pleural effusion
d. nephrotic syndrome
e. myxedema

237. Indications for hospital admission in patients with community-acquired pneumonia include
a. elderly patients (more than 65 years of age)
b. significant comorbidity
c. failure of outpatient therapy
d. inability to take oral medications
e. all of these

238. Which of the following is true about silicosis?
a. Mesothelioma is a complication of silicosis.
b. It is caused by exposure to asbestos fibers.
c. There is an increased risk of developing tuberculosis in patients with silicosis.
d. Silicosis generally presents as a pneumothorax on chest x-ray.
e. Patients with silicosis should not receive any antituberculous therapy.

239. Obstructive airway defect is characterized on pulmonary function testing by
a. reduced FEV_1/FVC ratio
b. decreased total lung capacity (TLC)
c. reduced residual volume (RV)
d. decreased residual volume/total lung capacity (RV/TLC)
e. decrease in diffusing capacity (DLCO)

240. Which of the following is the FIRST-LINE THERAPY in the management of an ACUTE asthma attack?
a. Steroids
b. β_2 Agonists
c. Theophylline
d. Antibiotics
e. Magnesium sulfate

241. Which of the following is NOT a cause of bronchiectasis?
a. Allergic bronchopulmonary aspergillosis
b. Cystic fibrosis
c. Bacterial infections
d. Kartagener's syndrome
e. Pulmonary embolism

242. Which is/are the risk factors for pulmonary thromboembolism?
a. Surgery
b. Trauma
c. Obesity
d. Immobilization
e. All of these

243. Acute respiratory distress syndrome (ARDS) is differentiated from acute lung injury (ALI) on the basis of
a. the presence of bilateral interstitial infiltrates on chest x-ray
b. the severity of hypoxemia with PaO_2/FiO_2 ratio of less than 200 mm Hg
c. increased pulmonary capillary wedge pressure
d. reduced compliance
e. none of these

244. Indication for lung transplant include all of the following EXCEPT
a. COPD
b. cystic fibrosis
c. lung cancer
d. idiopathic pulmonary fibrosis (IPF)
e. primary pulmonary hypertension

245. Which of the following is true about sarcoidosis?
a. Its etiology is unknown.
b. It is more common in females.
c. It is more common in blacks than in whites.
d. Glucocorticoids are an effective form of therapy.
e. All of these.

246. A 30-year-old man presents to the emergency room with shortness of breath and right-sided pleuritic chest pain. His chest x-ray in the emergency room is normal. Arterial blood gas is measured while the patient is breathing room air. The results show a pH of 7.48, P_{CO_2} of 35, Pa_{O_2} of 68, and an oxygen saturation of 92 percent. What is his A-a gradient?
a. 20
b. 30
c. 40
d. 50
e. 60

247. Which of the following eosinophilic pulmonary syndromes may present without any peripheral eosinophilia?
a. Loeffler's syndrome
b. Acute eosinophilic pneumonia
c. Chronic eosinophilic pneumonia
d. Allergic granulomatosis of Churg-Strauss syndrome
e. Hypereosinophilic syndrome

248. Exposure to asbestos fibers can cause
a. lung cancer
b. mesothelioma
c. pulmonary fibrosis
d. pleural plaques
e. all of these

249. In patients with COPD, long-term oxygen supplementation is prescribed, if Pa_{O_2} is
a. 55 mm Hg or below
b. 65 mm Hg
c. 70 mm Hg
d. 75 mm Hg
e. none of these

250. The most common cause of masses in the posterior mediastinum is
a. vascular
b. esophageal diverticula
c. neurogenic tumors
d. lymphomas
e. bronchogenic cysts

251. The risk factors for the development of obstructive sleep apnea (OSA) include
a. obesity
b. snoring
c. adenotonsillar hypertrophy
d. retrognathia and macroglossia
e. all of these

252. INH prophylaxis against tuberculosis is indicated in which of the following patients with a positive PPD?
a. HIV infected persons
b. Close contacts of tuberculosis patients
c. Recently infected persons
d. Persons with high risk medical conditions
e. All of these

253. The causes of hemoptysis include all of the following EXCEPT
a. bronchogenic carcinoma
b. acute bronchitis
c. Goodpasture's syndrome
d. emphysema
e. bronchiectasis

Pulmonary

Answers

234. The answer is d. *(Fauci, 14/e, pp 1466–1468.)* PPH has a female-to-male preponderance (1.7:1), with most patients presenting in the third and fourth decades. The etiology of PPH remains unknown, and other causes of pulmonary hypertension need to be ruled out before a diagnosis of PPH is made. The findings of pulmonary function tests are usually normal in PPH, although a mild restrictive pattern may be seen in some. COPD can cause secondary pulmonary hypertension but not PPH. Every patient with PPH should be given a trial of vasodilators to assess the response. Anticoagulation therapy also has been advocated based on the evidence that thrombosis in situ is common.

235. The answer is d. *(Fauci, 14/e, p 1473.)* Transudative and exudative effusions are distinguished by measuring the lactate dehydrogenase (LDH) and protein levels in the pleural fluid. Exudative pleural effusions meet at least one of the criteria as mentioned in choices A, B, and C, whereas transudative effusions meet none.

236. The answer is c. *(Fauci, 14/e, p 1475.)* The most common causes of transudative pleural effusions include CHF, cirrhosis, nephrotic syndrome, peritoneal dialysis, superior vena cava obstruction, myxedema, urinothorax, and pulmonary embolism. Of note, pulmonary embolism can cause both a transudative as well as an exudative effusion.

237. The answer is b. *(Fauci, 14/e, p 1441.)* Elderly patients and patients with other comorbid illnesses have a higher chance of complications following a community-acquired pneumonia. In addition to all of the other choices mentioned, patients with tachypnea, tachycardia, hypoxemia (Pa_{O_2} <60 mm Hg), hypotension, and acute alteration in mental status are also candidates for hospital admission.

238. The answer is c. *(Fauci, 14/e, p 1432.)* Workers exposed while sandblasting, tunneling through rock with high quartz content, or manufacturing of abrasive soaps can develop silicosis. Chest x-ray findings include a reticular pattern of irregular densities, mostly in the upper lung zones. The nodular fibrosis may be progressive with formation of irregular masses of greater than 1 cm each. These masses can become quite large and coalesce to form progressive massive fibrosis (PMF). Hilar lymph nodes may calcify in as few as 20 percent of cases and can produce the characteristic (e.g., shell) calcification. Patients with silicosis are at greater risk of acquiring *Mycobacterium tuberculosis* and atypical mycobacterial infections. Treatment or prophylaxis for tuberculosis is indicated for patients with silicosis and a positive tuberculin test.

239. The answer is a. *(Fauci, 14/e, p 1412.)* Pulmonary function tests are done mainly to distinguish between obstructive and restrictive defects. The hallmark of obstructive defect is a decrease in the expiratory flow rate, as manifested by a decrease in FEV_1/FVC ratio. TLC is normal or increased. RV is elevated owing to the air trapping during expiration. This results in an increase of RV/TLC ratio. Vital capacity is frequently decreased in obstructive defects because of striking elevations in RV with only minor changes in TLC.

240. The answer is b. *(Fauci, 14/e, p 1425.)* The most effective treatment for acute episodes of asthma is administration of aerosolized β_2 agonists. In emergency situations, they can be given every 20 min until the attack has subsided or the patient develops any side effects. Thereafter, the frequency can be reduced to every 2 to 4 h until the attack has totally subsided. Other drugs have some role in asthma but are not the first-line therapeutic agents.

241. The answer is e. *(Fauci, 14/e, p 1446.)* Bronchiectasis is a consequence of inflammation and destruction of the bronchial walls. Infectious causes include pneumonia with virulent organisms (e.g., *Pseudomonas* and *Staphylococcus aureus*). Noninfectious causes include immune-mediated inflammation (e.g., in allergic bronchopulmonary aspergillosis). In yellow-nail syndrome, which is due to hypoplastic lymphatics, the triad of lymphedema, pleural effusion, and yellow discoloration of the nails is accompanied by bronchiectasis in approximately 40 percent of the patients. Patients with Kartagener's syndrome have ciliary motility defect. The syndrome consists of situs inversus accompanied by bronchiectasis and sinusitis. Pulmonary embolism has not been reported to cause bronchiectasis.

242. The answer is e. *(Fauci, 14/e, p 1469)* The majority of the pulmonary emboli arise from the lower extremities. Any insult or risk factors that can cause stasis of blood in the lower extremities, and hence thrombus formation, predisposes the clot to dislodge into the pulmonary circulation, thereby causing pulmonary embolism. Local trauma, hypercoagulability, and stasis are the main factors causing thromboembolic disease. However, many patients can have an underlying inherited predisposition that remains clinically silent until an acquired stress, such as surgery, obesity, or pregnancy.

243. The answer is b. *(Fauci, 14/e, p 1483.)* ARDS and ALI are both characterized by bilateral interstitial infiltrates, normal pulmonary capillary wedge pressure, and low compliance. However, the severity of hypoxemia, as defined by the Pa_{O_2}/Fi_{O_2} ratio distinguishes the two syndromes from each other. In ARDS, patients have a refractory hypoxemia with a Pa_{O_2}/Fi_{O_2} ratio of less than 200 mm Hg whereas, in patients with ALI, this ratio is greater than 200.

244. The answer is c. *(Fauci, 14/e, p 1491.)* COPD accounts for about 60 percent of all single-lung transplants and about 30 percent of bilateral lung transplants. Cystic fibrosis accounts for approximately 36 percent of bilateral lung transplants, and the rest of the lung transplants are due to miscellaneous reasons, including idiopathic pulmonary fibrosis, and PPH, among others. Cancer in the lungs or outside of the lungs would preclude patients from undergoing lung transplantation.

245. The answer is e. *(Fauci, 14/e, p 1922.)* Sarcoidosis is a chronic, multisystem disorder of unknown cause. It affects both sexes, although females appear to be slightly more susceptible than males. There is a remarkable diversity of the prevalence of sarcoidosis among certain ethnic and racial groups. The prevalence is from 10 to 40 per 100,000 in the United States, where the majority of patients are black, with a ratio of blacks to whites ranging from 10:1 to 17:1. The therapy of choice for sarcoidosis is oral glucocorticoids. The disease responds well to steroids, but the treatment for symptomatic sarcoid patients generally lasts months.

246. The answer is d. *(Fauci, 14/e, p 1415.)* The patient most likely has pulmonary embolism. A useful calculation is the assessment of alveolar oxygenation and calculation of the gradient between alveolar and arterial partial pressures of the oxygen. At room air, the Pa_{O_2} (alveolar) can be calculated by $Pa_{O_2} = 150 - 1.25 \times Pa_{CO_2}$. Once Pa_{O_2} is determined, the A–a gradient is simply the difference between Pa_{O_2} and arterial Pa_{O_2}. In a healthy young person breathing room air, the $Pa_{O_2} - Pa_{O_2}$ is normally less than 15 mm Hg; this value increases with age and may be as high as 30 mm Hg in elderly patients.

247. The answer is b. (*Fauci, 14/e, p 1429.*) The group of idiopathic eosinophilic pneumonia consists of diseases of varying severity. *Loeffler's syndrome* was originally reported as migratory pulmonary infiltrates that, in some patients, may be secondary to parasites or drugs. *Acute eosinophilic pneumonia* is a recently described syndrome characterized by an acute febrile illness of less than 7 days in duration and may or may not present with peripheral eosinophilia. *Chronic eosinophilic syndrome* presents with significant systemic symptoms of weeks or months in duration and presents with peripheral eosinophilia. *Allergic angiitis and granulomatosis* of Churg-Strauss syndrome is a multisystem vasculitic disorder that frequently involves skin, kidneys, and nervous system in addition to the lungs. This is also manifested by peripheral eosinophilia. The *hypereosinophilic syndrome* is characterized by the presence of over 1500 eosinophils per micoliter of peripheral blood for 6 months or longer.

248. The answer is e. (*Fauci, 14/e, p 1431.*) Asbestos exposure can cause lung cancer, and concomitant cigarette smoking potentiates the effect. There is a minimum lapse of 15 to 19 years between the first exposure and the development of the disease. Mesothelioma is a pleural-based tumor, usually occurring after 30 to 35 years of initial exposure. This tumor is not related to smoking. Pleural fibrosis is slowly evolving parenchymal scarring. Usually there is a time lapse of about 10 years between the first exposure and the development of asbestosis. Pleural plaques are the most benign form of abnormality caused by asbestos exposure and generally are asymptomatic.

249. The answer is a. (*Fauci, 14/e, p 1457.*) If Pa_{O_2} is persistently below 55 mm Hg, supplemental oxygen should be prescribed. However, if room air Pa_{O_2} is between 55 and 60 mm Hg, supplemental oxygen may still be prescribed if the patient has signs of cor pulmonale or secondary erythrocytosis or signs of right heart failure. In patients with severe hypoxemia, supplemental oxygen improves exercise tolerance and neurophysiologic functions and alleviates pulmonary hypertension. It tends to improve survival if used more than 15 to 19 h a day.

250. The answer is c. (*Fauci, 14/e, p 1475.*) The most common tumor in the posterior mediastinum is the neurogenic tumor. Other masses found in the posterior mediastinum are meningoceles, gastroenteric cysts, and esophageal diverticula. The most common masses in the middle mediastinum are vascular masses, lymph node enlargement from metastases or granulomatous disease, and pleuropericardial and bronchogenic cysts. In the anterior mediastinum, the most common lesions are thymomas, lymphomas, teratomas, and thyroid masses.

251. The answer is e. (*Fauci, 14/e, 1480.*) Pathophysiologically, any condition causing narrowing of the upper airway may result in OSA. Obesity contributes to OSA by increasing fat deposition in the soft tissues of the pharynx or by compressing the pharynx by superficial fat masses in the neck. In most patients, snoring antedates the development of obstructive events by many years. However, not every snorer would have OSA. In a minority of patients, a structural compromise, such as adenotonsillar hypertrophy, retrognathia, or macroglossia, can also contribute to the development of OSA.

252. The answer is e. (*Fauci, 14/e, 1013.*) Persons at risk for developing tuberculosis include close contacts, those infected with HIV, and patients who are recent converters. Patients with a high-risk medical condition include those with diabetes mellitus or who are on prolonged therapy with glucocorticoids, and others on immunosuppressive therapy or with end-stage renal disease.

253. The answer is d. (*Fauci, 14/e, pp 196–197.*) Hemoptysis can occur because of a tracheobronchial source, pulmonary parenchymal source, primary vascular disease, coagulopathy, or immune-mediated disease (e.g., Goodpasture's syndrome). The most common cause of mild hemoptysis in the United States is acute bronchitis. Hemoptysis is not a feature of emphysema and, in patients with emphysema who present with hemoptysis, another cause should be investigated.

RENAL/NEPHROLOGY

Questions

DIRECTIONS: Each item below contains a question or incomplete statement followed by suggested responses. Select the **one best** response or the **matching** response to each question.

254. Which finding is not common in acute glomerulonephritis?
a. Hypertension
b. Red cell casts in the urine
c. Peripheral eosinophilia
d. Vascular congestion
e. Oliguria

255. Which finding is fairly specific for chronic renal failure?
a. Anemia
b. Hyaline casts
c. Broad casts in urinalysis
d. Proteinuria
e. Hypocalcemia

256. Sterile pyuria is not a common finding in
a. chronic interstitial nephritis
b. renal transplant rejection
c. tuberculous infection
d. prostatic hypertrophy

257. Which finding is not found in atheroembolic acute renal failure?
a. Acute flank pain
b. Livedo reticularis
c. Arteriolar plaques in the retina
d. Digital ischemia
e. Eosinophilia

258. A high fractional excretion of sodium is typically found in
a. heart failure
b. urinary tract obstruction
c. acute tubular necrosis
d. acute glomerulonephritis
e. hepatorenal syndrome

259. In chronic renal failure, which of the following is not a typical finding?
a. Hyperkalemia
b. Anemia
c. Hypercalcemia
d. Prolonged bleeding time
e. Hyperparathyroidism

260. Which of the following statements is true in the management of acute renal failure?
a. Metabolic acidosis is fully corrected with bicarbonate.
b. Hyperphosphatemia is primarily managed with dialysis.
c. Low-dose dopamine is used to shorten the duration of renal failure.
d. Hypervolemia is managed with high-dose loop diuretics.
e. Hyponatremia is corrected by administration of sodium salts.

261. Which of the following describes bone abnormalities in patients with chronic renal failure?
a. Osteitis fibrosis cystica is a result of oversuppression of parathyroid hormone (PTH).
b. Adynamic bone disease is associated with myopathy.
c. Osteomalacia is due to excessive accumulation of magnesium.
d. Hyperparathyroidism responds well to 1,25-dihydroxyvitamin D.
e. Amyloidosis in dialysis patients is similar in etiology to patients who are not on dialysis.

262. Which of the following statements is true regarding hematologic disorders in CRF?
a. Resistance to erythropoietin is most commonly due to aluminum overload.
b. Erythropoietin is associated with worsening hypertension.
c. The major cause of death in CRF is sepsis.
d. Abnormal bleeding responds best to platelet transfusion.
e. Leukocyte function is generally unimpaired.

263. Which of the following is useful in retarding progression of renal failure?
a. Aggressive blood pressure control
b. Decrease in protein intake
c. Angiotensin-converting enzyme (ACE) inhibitors more than other antihypertensive drugs
d. Erythropoietin for anemia

264. Rapidly progressive glomerulonephritis is characterized by the following EXCEPT
a. red cell casts in the urine
b. oliguria
c. hypertension
d. nephrotic syndrome
e. edema

265. Which of the following serologic findings is associated with linear staining of the glomerulus on immunofluorescence?
a. Anti-GBM (glomerular basement membrane) antibody
b. Low-complement immune-complex glomerulonephritis
c. Antineutrophil cytoplasmic antibody (ANCA)-associated renal disease
d. Membranoproliferative glomerulonephritis

266. ANCA is typically present in which systemic disease?
a. Goodpasture's syndrome
b. Wegener's granulomatosis
c. Systemic lupus erythematosus (SLE)
d. Thrombotic thrombocytopenic purpura (TTP)

267. Postinfectious glomerulonephritis is characterized by which of the following?
a. Most cases in an epidemic are subclinical.
b. Hematuria typically develops within a week of infection.
c. It is more common with pharyngeal than cutaneous streptococcus infection.
d. Focal proliferative glomerulonephritis is seen on renal biopsy.
e. Children are often left with residual renal impairment.

268. Which of the following is true of Goodpasture's syndrome?
a. The clinical presentation is largely the same in different age groups.
b. The target antigen in the glomerulus is elastin.
c. Complement levels are typically normal.
d. Plasmapharesis enables dialysis-dependent patients to recover renal function.
e. Transplantation is contraindicated because of disease recurrence.

269. Which of the following is not a finding typical of nephrotic syndrome?
a. Increased susceptibility to infection
b. Hypocalcemia
c. Hyperlipidemia
d. Increased tendency to bleed

270. Minimal change disease
a. is the most common cause of nephrotic syndrome in adults
b. has characteristic abnormalities on light microscopy
c. is a common cause of progressive renal insufficiency
d. may be caused by intake of nonsteroidal drugs
e. is usually treated with a combination of steroids and cytotoxic agents

271. Comparing focal sclerosis and membranous glomerulopathy, which is NOT true?
a. Both can be idiopathic or secondary.
b. Focal sclerosis has a better long-term prognosis and responds better to steroids.
c. Malignancies are associated with membranous glomerulopathy.
d. The lesion in HIV is most commonly focal sclerosis.
e. Focal sclerosis is a common response to previous nephron loss or injury.

272. Which is NOT true of membranoproliferative glomerulonephritis?
a. It includes depressed complement levels.
b. It is associated with hepatitis C infection.
c. It is an immune-complex-mediated disease.
d. Type II is associated with C3 nephritis factor.
e. It responds to steroid therapy.

273. Which may cause acute renal failure in patients with nephrotic syndrome?
a. Dietary protein restriction
b. ACE inhibitors
c. Lipid-lowering agents
d. Loop diuretics

274. Which statement is NOT true of diabetic nephropathy?
a. It is the most common cause of renal failure in United States.
b. Microalbuminuria is the first manifestation of the disease.
c. It is almost always associated with retinopathy in insulin-dependent diabetes mellitus (IDDM).
d. ACE inhibitors are indicated only for patients with hypertension.
e. Hyperkalemia and type IV renal tubular acidosis (RTA) are commonly found.

275. Which disease presents with predominantly tubulointerstitial involvement?
a. Systemic lupus erythematosus
b. Sjögren's syndrome
c. Rheumatoid arthritis
d. Essential mixed cryoglubulinemia

276. Which disease presents with predominantly glomerular involvement?
a. Analgesic nephropathy
b. Uric acid nephropathy
c. Lead nephropathy
d. Light-chain deposition disease

277. Which of the following is a common cause of isolated hematuria?
a. Alport's syndrome
b. Thin-basement-membrane disease
c. Amyloidosis
d. IgA nephropathy

Items 278–281

Match the type of RTA in A–C:
a. Type I RTA
b. Type II RTA
c. Type IV RTA

278. Hyperparathyroidism and nephrocalcinosis

279. Urine pH always greater than 5.5

280. Nonsteroidal anti-inflammatory agents

281. Fanconi's syndrome

Items 282–285

Match the type of inherited kidney disease in A–D:
a. Autosomal dominant polycystic kidney disease
b. Autosomal recessive polycystic kidney disease
c. Medullary sponge kidney
d. Juvenile nephronophthisis/medullary cystic disease

282. Both hepatic fibrosis and renal failure in children

283. Small kidneys with interstitial fibrosis and renal failure

284. Nephrocalcinosis, hematuria, and urinary tract infection

285. Hepatic cysts, intracranial aneurysm, and colonic diverticuli

Items 286–289

Match the laboratory findings in A–D:

	pH	pCO_2	HCO_3	anion gap
a.	7.3	30	15	10
b.	7.38	55	30	12
c.	7.26	60	26	10
d.	7.49	47	35	15

286. Vomiting

287. COPD

288. Diarrhea

289. Respiratory arrest

Items 290–293

Match the type of hyponatremia in A–D:
a. Hypervolemic hyponatremia
b. Isovolemic hyponatremia
c. Hyponatremic hyponatremia
d. Pseudohyponatremia

290. Addison's disease

291. Nephrotic syndrome

292. Hypothyroidism

293. Hypertriglyceridemia

Items 294–298

Match the condition in A–E:
a. Hyperaldosteronism
b. Bartter's syndrome
c. Scleroderma
d. Renal artery stenosis
e. Pheochromocytoma

294. Hypertension with microangiopathic anemia

295. Hypertension with elevated renin level

296. Hypertension with metabolic alkalosis

297. Hypertension with tachycardia

298. Normotension with metabolic alkalosis

Renal/Nephrology

Answers

254. The answer is c. *(Fauci, 14/e, p 1495.)* Peripheral eosinophilia is usually seen in acute interstitial nephritis. The other findings are commonly seen in glomerulonephritis.

255. The answer is c. *(Fauci, 14/e, p 1496.)* The finding of broad casts reflects compensatory dilatation of surviving nephrons. Hyaline casts are a nonspecific finding. Proteinuria can be present in various stages of renal disease. Anemia and hypocalcemia can be present in acute renal failure and are usually multifactorial.

256. The answer is d. *(Fauci, 14/e, p 1497.)* Prostatitis, and not prostatic enlargement alone, causes leukocytes to appear in the urine. Tuberculous infections may be difficult to prove in routine media. In renal transplant rejection and chronic interstitial nephritis, infiltrating leukocytes appear in the urine.

257. The answer is a. *(Fauci, 14/e, p 1508.)* Flank pain can be caused by renal infarction through thromboembolism of a major renal vessel. Atheroembolism reflects small vessel disease, and suspicion of it is usually aroused by a physical examination. Eosinophilia is not specific for atheroembolism, because it can be seen in acute interstitial nephritis.

258. The answer is c. *(Fauci, 14/e, p 1508.)* In acute tubular necrosis, tubular damage prevents reabsorption of filtered sodium. In the other disorders listed, renal hypoperfusion causes avid sodium retention. Creatinine is not reabsorbed, hence leading to a low fractional excretion of sodium.

259. The answer is c. *(Fauci, 14/e, p 1510.)* In chronic renal failure, 1,25-dihydroxyvitamin D production is impaired and hypocalcemia results. Anemia results from decreased erythropoietin production. Hyperphosphatemia and hyperkalemia result from decreased excretion, and prolonged bleeding results from retained uremic toxins.

260. The answer is d. *(Fauci, 14/e, p 1512.)* Metabolic acidosis is usually corrected if the pH is less than 7.2. Hyperphosphatemia is managed initially with phosphate binders. Dopamine has not been shown to impact renal recovery. Hyponatremia is managed with water restriction. Hypervolemia is managed by decreasing salt and water intake and loop diuretics.

261. The answer is d. *(Fauci, 14/e, p 1516.)* Osteitis fibrosa is due to hyperparathyroidism and is associated with myopathy. Adynamic bone disease is associated with oversuppression of PTH. Osteomalacia is due to excessive aluminum accumulation. Amyloidosis in dialysis patients is due to β2 microglobulin and not the amyloid proteins seen in usual amyloidosis. 1,25-Dihydoxyvitamin D suppresses PTH production.

262. The answer is b. *(Fauci, 14/e, p 1518.)* Resistance to erythropoietin is most commonly due to iron deficiency despite oral iron intake. In a third of patients, the rise in hematocrit with erythropoietin worsens hypertension. Abnormal bleeding is treated with intensive dialysis, vasopressin, and estrogens among other measures but not with platelets. Although leukocyte function is impaired, sepsis is the second leading cause of death after cardiovascular disease.

263. The answer is d. *(Fauci, 14/e, p 1519.)* Erythropoietin is used to correct anemia of chronic renal failure. Most evidence suggests that it does not hasten the decline of renal function but is not otherwise protective. The other measures are recommended, although protein restriction should be carefully monitored to avoid malnutrition.

264. The answer is d. *(Fauci, 14/e, p 1536.)* Nephrotic range proteinuria is uncommon in rapidly progressive glomerulonephritis. The other components are typical findings of the so-called nephritic syndrome.

265. The answer is a. *(Fauci, 14/e, p 1537.)* Anti-GBM disease is characterized by linear staining of the basement membrane. Immune-complex disease, such as lupus or membranoproliferative glomerulonephritis, is characterized by granular staining. ANCA-associated renal diseases were formerly called pauci-immune due to absence of staining by immunofluorescence.

266. The answer is b. *(Fauci, 14/e, p 1537.)* Goodpasture's syndrome consists of pulmonary hemorrhage and renal failure. The latter alone is called anti-GBM disease. ANCA is typically present in vasculitic disorders, such as Wegener's granulomatosis, microscopic polyangiitis, and Churg-Strauss syndrome. SLE may have ANCA positivity if vasculitis is a prominent feature. TTP is primarily due to endothelial injury.

267. The answer is a. *(Fauci, 14/e, p 1539.)* Most cases of postinfectious glomerulonephritis are subclinical and found among contacts of the index case. Hematuria typically occurs 10 days or more after the infection in comparison to IgA nephropathy, in which the hematuria follows very closely thereafter. Unlike rheumatic fever, it is more common after cutaneous infection, and the lesion is a diffuse proliferative glomerulonephritis. The prognosis in children is usually excellent.

268. The answer is c. *(Fauci, 14/e, p 1539.)* Anti-GBM disease in the older age group is characterized by absence of pulmonary hemorrhage, as compared with the younger age group. The antigen is the NC1 domain of α3 chain of type IV collagen. Although plasmapharesis is the treatment of choice, dialysis-dependent patients rarely recover renal function. Transplantation can be carried out after seronegativity for a defined period, with disease recurring only rarely.

269. The answer is d. *(Fauci, 14/e, p 1540.)* Nephrotic syndrome confers increased susceptibility to infection due to loss of immune globulins. Hypocalcemia is due to vitamin D deficiency. Hypercoagulability is typically present.

270. The answer is d. *(Fauci, 14/e, p 1541.)* Minimal change disease is the most common cause of nephrotic syndrome in children who are typically treated without biopsy. As the name implies, the glomeruli are largely normal under light microscopy and foot-process effacement is seen on electron microscopy. The prognosis is excellent in children. Some renal insufficiency may be present in a few adults. Nonsteroidal drugs can lead to nephrotic syndrome by causing a similar histologic appearance. The majority of children and half of adults will respond to prednisone alone.

98   Pathophysiology

271. The answer is b. *(Fauci, 14/e, p 1542.)* Focal sclerosis responds poorly to steroids and can lead to progressive renal failure. It can be idiopathic or associated with obesity, HIV, and previous nephron loss due to lupus nephritis or reflux nephropathy. Membranous nephropathy may be idiopathic, a manifestation of collagen vascular disease, drug toxicity, or underlying malignancy.

272. The answer is e. *(Fauci, 14/e, p 1543.)* Membranoproliferative glomerulonephritis can be associated with different disorders and is not generally responsive to steroid therapy.

273. The answer is d. *(Fauci, 14/e, p 1544.)* High-dose diuretics can precipitate acute renal failure in nephrotics, because the intravascular volume cannot be defended by the hypoalbuminemia. The other three measures are generally considered helpful in nephrosis, although dietary protein restriction needs monitoring to guard against malnutrition.

274. The answer is d. *(Fauci, 14/e, p 1546.)* Microalbuminuria precedes macroalbuminuria by 5 to 10 years. The incidence of retinopathy is lower in non-IDDM compared with IDDM, where it is nearly universal. ACE inhibitors, which are now being given to most diabetics with microalbuminuria, are beneficial for reducing protein excretion and preserving renal function.

275. The answer is b. *(Fauci, 14/e, p 1549.)* Sjögren's syndrome presents as a tubulointerstitial nephritis. Lupus may present with proliferative or membranous changes along with interstitial involvement. Rheumatoid arthritis usually presents with amyloidosis, whereas cryoglobulinemic nephropathy is a membranoproliferative lesion.

276. The answer is d. *(Fauci, 14/e, p 1555.)* Light-chain deposition disease, found in paraproteinemias, presents with proteinuria and nephrotic syndrome due to glomerular deposition. The other diseases are examples of tubulointerstitial renal disease characterized by low albumin excretion in urine.

277. The answer is c. *(Fauci, 14/e, p 1544.)* Amyloidosis presents with nephrotic syndrome and systemic involvement. The other diseases present with isolated hematuria and may (as in Alport's syndrome and IgA nephropathy) or may not (as in thin-basement-membrane disease) progress to renal failure.

Renal/Nephrology Answers

278–281. The answers are a, a, c, b. *(Fauci, 14/e, pp 1566–1567.)* Type I RTA, which is due to a loss of acidification in the distal tubule, always presents with an elevated pH. Type II RTA, which is due to proximal bicarbonate loss, may have a urine pH of less than 5.5 if serum bicarbonate falls within the reabsorptive threshold. Type IV RTA, which is due to a defect in ammonia production and to hypoaldosteronism, also has a pH of less than 5.5.

282–285. The answers are b, d, c, a. *(Fauci, 14/e, p 1563.)* The different inherited cystic diseases have varying presentations and modes of inheritance. Medullary sponge kidney does not lead to renal failure like the other three. Polycystics have large kidneys, medullary sponge kidney is usually normal, and medullary cystic disease kidney is usually small at time of discovery.

286–289. The answers are d, b, a, c. *(Fauci, 14/e, p 279.)* Vomiting causes a metabolic alkalosis. COPD is a chronic respiratory acidosis with metabolic compensation, whereas a respiratory arrest presents without metabolic adjustment and hence the serum bicarbonate is lower and so is the pH. Diarrhea is a metabolic acidosis with a normal anion gap.

290–293. The answers are c, a, c, d. *(Fauci, 14/e, p 269.)* Hyponatremia is best approached by categorizing it into different volume states. This can be differentiated by good history, physical exam, and the urinary indices.

294–298. The answers are c, d, a, e, b. *(Fauci 14/e, p 1560.)* Secondary hypertension can be due to varying causes, some of which are enumerated here. Clues to a secondary cause include young or advanced age at onset of hypertension, hard-to-control hypertension, and abnormalities in laboratory examination findings.

GASTROENTEROLOGY

Questions

DIRECTIONS: Each item below contains a question or incomplete statement followed by suggested responses. Select the **one best** response, the **matching** response, or the **true/false** response to each question.

299. All of the following may present with hematemesis EXCEPT
a. Mallory-Weiss tear
b. peptic ulcers
c. gastric cancer
d. esophagitis
e. diverticulosis

300. Bleeding peptic ulcer disease may present with
a. vomiting of coffee-ground emesis
b. black stool
c. vomiting of red blood
d. none of these
e. all of these

Items 301–305

Biopsies should routinely be performed based on the following findings noted on upper endoscopy (answer True or False for each):

301. Gastric ulcer

302. Duodenal ulcer

303. Arteriovenous malformation (AVM)

304. Esophageal ulcer

305. Lower esophageal sphincter

306. Hepatitis C infection can be transmitted in all of the following manners EXCEPT by
a. the fecal/oral route
b. intravenous drug use
c. promiscuous sex
d. tattoos
e. blood transfusions

307. The percentage of patients with acute hepatitis C that go on to have chronic disease is
a. 5 to 10
b. 15 to 20
c. 25 to 30
d. 40 to 50
e. 60 to 70

308. The extent of liver damage done by chronic hepatitis B or C infection can best be gauged by evaluating
a. symptoms
b. elevation of serum transaminases
c. duration of infection
d. liver biopsy

309. Symptoms of acute pancreatitis may include
a. nausea and vomiting
b. abdominal pain
c. low-grade fever
d. elevated amylase
e. all of these

310. Chronic pancreatitis may be reliably diagnosed in a patient presenting with
a. jaundice
b. abdominal pain
c. diarrhea
d. nausea and vomiting
e. calcified pancreas on flat plate x-ray of the abdomen

311. The ideology of acute pancreatitis includes all of the following EXCEPT
a. gallstones
b. alcohol
c. trauma
d. viral illness
e. diabetes

312. All of the following are accurate regarding therapeutic endoscopic retrograde cholangiopancreatography (ERCP) EXCEPT
a. It allows removal of gallstones.
b. It allows placement of stents.
c. Is indicated initially for severe gallstone pancreatitis.
d. It can nonsurgically relieve obstructive jaundice.
e. It results in complications in about 20 percent of cases.

313. All of the following can be diagnosed using esophageal manometry EXCEPT
a. achalasia
b. hypertensive lower esophageal sphincter (LES)
c. esophageal spasm
d. nonspecific esophageal motility disorders
e. obstructive causes of dysphagia

Items 314–316

Match each patient with the correct diagnosis:.
a. Achalasia
b. Esophageal spasm
c. Hypertensive LES

314. A 60-year-old man with dysphagia, esophageal motility, and absence of peristalsis in his body, with a high-pressure LES that relaxes incompletely

315. A 45-year-old woman with chest pain for which cardiac cause has been ruled out and whose esophageal motility shows pressure waves of a very high amplitude lasting for 2 to 3 s

316. A 40-year-old man with occasional dysphagia, who otherwise is well, whose esophageal motility shows a LES amplitude of approximately 60 mmHg that relaxes completely with swallowing

317. The presence of gastroesophageal reflux is best diagnosed by
a. computed tomographic (CT) scan of the chest
b. physical exam
c. laboratory evaluation
d. barium swallow
e. medical history

318. The most common location for a gastric ulcer is the
a. fundus
b. greater curve
c. cardia
d. body
e. antrum

Items 319–322

For each finding, choose the peptic ulcer disease most likely to be appropriate:
a. Gastric ulcer
b. Duodenal ulcer
c. Both
d. Neither

319. *Helicobacter pylori* is a causative agent.

320. Most likely to be caused by nonsteroidal use.

321. Surgery commonly used for symptomatic control.

322. Very low risk for malignancy.

Items 323–327

Which of the following statements regarding *Helicobacter pylori* are true? Answer True or False for each:

323. It is almost universally associated with antral gastritis.

324. It is a frequent cause of both gastric and duodenal ulcers.

325. It may cause epigastric symptoms of dyspepsia and functional pain in adults.

326. It is infrequently seen in the population as a whole.

327. It can be effectively eradicated with medication.

328. All of the following are consistent with the diagnosis of malabsorption EXCEPT
a. steatorrhea
b. villous atrophy on small intestinal biopsy
c. low calcium levels
d. decreased iron levels
e. elevated zinc levels

329. Causes of malabsorption include all of the following EXCEPT
a. chronic pancreatitis
b. Crohn's disease
c. gastric surgery
d. small bowel ischemia
e. sigmoid resection

330. All of the following are consistent with a diagnosis of ulcerative colitis EXCEPT
a. presentation with bloody diarrhea
b. increased risk of colorectal cancer
c. continuous colonic involvement
d. granulomas present on colon biopsy

331. Crohn's disease can be distinguished from ulcerative colitis by
a. frequent involvement of the small bowel as well as colon
b. skip lesions noted in the gastrointestinal tract
c. presentation with nausea and vomiting secondary to obstruction
d. none of these
e. all of these

Items 332-335

Choose the best initial imaging study in the following situations:
a. Ultrasound
b. CT scan
c. Magnetic resonance imaging (MRI)
d. Barium swallow

332. Suspected gallstone disease

333. Evaluation of obstructive jaundice

334. Suspicion of metastatic disease to the liver with equivocal computed axial tomography (CAT) scan

335. Evaluation of dysphagia

336. Symptoms due to *Clostridium difficile* infection can be accurately diagnosed by the
a. presence of diarrhea
b. stool positive for WBCs
c. history of recent antibiotic usage
d. pseudomembranes noted on a sigmoidoscopy

337. Which would characterize the frequency of chronic disease following acute hepatitis A infection?
a. Rare
b. Infrequent
c. Common
d. Typical
e. Nonexistent

338. Which of the laboratory patterns is most consistent with the diagnosis of hemochromatosis?
a. Low iron, low TIBC (total iron-binding capacity), and low ferritin
b. Low iron, low TIBC, and increased ferritin
c. Low iron, increased TIBC, and decreased ferritin
d. Increased iron, increased TIBC, and increased ferritin

339. Which of the following is seen most commonly in association with primary biliary cirrhosis (PBC)?
a. Positive antinuclear antibody (ANA)
b. Increased ceruloplasmin
c. Increased ferritin
d. Positive hepatitis B surface antigen
e. Positive antimitochondrial antibody (AMA)

340. All of the following are potential consequences of chronic hepatitis B or C infection EXCEPT
a. hepatoma
b. cirrhosis
c. liver failure
d. inability to be gainfully employed
e. hepatic adenoma

341. All of the following may be seen as a presentation of acute viral hepatitis EXCEPT
a. fatigue
b. orange/brown urine
c. anorexia
d. alanine aminotransferase (ALT) level of 500 (normal, 45)
e. spider angiomas

342. Protective vaccines are available for which of the following hepatitis viruses?
a. B and C
b. A and C
c. C and D
d. A and D
e. A and B

343. Liver biopsy done in a patient suspected of having acute alcohol hepatitis may reveal the following EXCEPT
a. steatosis
b. Mallory bodies
c. polymorphs
d. ballooning degeneration
e. all of these

GASTROENTEROLOGY

Answers

299. The answer is e. *(Fauci, 14/e, p 246.)* Hematemesis or vomiting of blood represents an upper gastrointestinal (GI) source of blood loss; diverticulosis represents a lower GI source of blood loss.

300. The answer is e. *(Fauci, 14/e, p 1607.)* Vomiting of blood or hematemesis can be either fresh blood or dark blood (e.g., coffee grounds). Black blood is melena and is typical for an upper GI bleed.

301–305. The answers are t, f, f, t, f. *(Fauci, 14/e, pp 568, 1607.)* Gastric ulcers, particularly large ones, should be biopsied to rule out any malignancy. Due to the low incidence of malignancy in duodenal ulcers, biopsies are rarely done. AVMs are recognized visually and no biopsies are necessary, particularly with the increased incidence of bleeding. Esophageal ulcerations are likewise biopsied to rule out any malignancy. Biopsy specimens are not routinely taken of the lower esophageal sphincter area unless mucosal abnormalities are suspected.

306. The answer is a. *(Fauci, 14/e, p 1686.)* Hepatitis C requires parenteral exposure for transmission. It is most commonly spread through intravenous drug abuse and promiscuity.

307. The answer is e. *(Fauci, 14/e, p 1690.)* Up to 75 percent of hepatitis C patients may go on to become carriers and have chronic disease. Given the lack of effective current therapy, it makes prevention all the more important.

308. The answer is d. *(Fauci, 14/e, p 1700.)* Obtaining a piece of liver tissue is the only way to judge the histologic damage caused by chronic viral infection. There is a poor correlation between symptoms and transaminase elevation, as well as duration of infection and liver damage.

309. The answer is e. *(Fauci, 14/e, p 1581.)* All are characteristics of acute inflammatory disease of the pancreas. The pain is primarily epigastric with radiation into the back and is usually a continuous, boring pain. Fever may or may not be present. Amylase and lipase are usually elevated acutely.

310. The answer is e. *(Fauci, 14/e, p 1581.)* Jaundice, abdominal pain, diarrhea and nausea, and vomiting all can be seen in patients with chronic pancreatitis but are nonspecific. The finding of calcification on KUB (kidneys, ureters, and bladder) x-rays is diagnostic for chronic pancreatitis.

311. The answer is e. *(Fauci, 14/e, p 1741.)* Gallstones and alcohol-induced disease account for 80 to 90 percent of the cases of acute pancreatitis. Trauma, viral illnesses, hypercalcemia, and medications are other less common causes; diabetes is not a cause of acute pancreatitis.

312. The answer is e. *(Fauci, 14/e, p 1584.)* Therapeutic ERCP is used most commonly for complicated gallstone disease, as well as for an evaluation of obstructive jaundice. When needed, sphincterotomy can be done or stents can usually be placed very safely, with an approximately 5 percent complication rate. The majority of complications are not serious.

313. The answer is e. *(Fauci, 14/e, p 1501.)* Achalasia shows a high-pressure nonrelaxing LES with absent motility in the body of the esophagus. Hypertensive LES has a high-pressure reading at the LES, and esophageal spasm shows high amplitude prolonged pressure waves. Nonspecific disorders include repetitive swallows and dropped waves that do not move through the whole esophagus. Obstructive causes of dysphasia are best diagnosed on barium swallow or endoscopy.

314–316. The answers are a, b, c. *(Fauci, 14/e, pp 1590–1591.)* Patients with achalasia present with a history of dysphagia or sensation of food sticking, which includes both solids and liquids. They have a long history of these symptoms and might come for treatment at an older age. The patients with esophageal spasm and hypertensive LES usually present at a younger age. Patients with spasm have more severe pain and their symptoms often are confused with a cardiac etiology. Each of these have characteristic motility findings.

317. The answer is e. *(Fauci, 14/e, p 1593.)* The diagnosis of gastroesophageal reflux is best diagnosed by history with the typical presentation of retrosternal burning, usually postprandially and sometimes nocturnally. It may be exacerbated by certain foods. In simple reflux disease, the results of a physical exam and the lab tests are unremarkable. CT findings of the chest are normal in reflux disease. Gastroesophageal reflux can be elicited on barium swallow but oftentimes is only an incidental and unrelated finding in patients who have no heartburn symptoms.

318. The answer is e. *(Fauci, 14/e, p 1605.)* Eighty-five to ninety percent of ulcers are found in the prepyloric and antral areas. Finding an ulcer in a different location is unusual but is not indicative of a higher incidence of malignancy.

319–322. The answers are c, a, d, b. *(Fauci, 14/e, p 1605.)* Helicobacter pylori is associated as a cause of approximately 95 percent of duodenal ulcers and 75 percent of gastric ulcers. Nonsteroidals primarily cause gastric pathology but can also cause duodenal disease as well. Surgery is rarely indicated for the control of symptoms with the medications currently available. There is a concern for malignancy in gastric ulcers and biopsies are performed because of this. Duodenal ulcers are at exceedingly low risk for any malignancy and therefore biopsies are not routinely performed.

323–327. The answers are t, t, f, f, t. *(Fauci, 14/e, p 1605.)* Nearly all patients infected with *H. pylori* have antral gastritis. It is associated as a causative agent in approximately 90 to 95 percent of duodenal ulcer and 75 percent of gastric ulcer patients. In adults, it rarely causes symptoms. It is frequent in the population, being present in approximately half of the adults in the United States, and its frequency is increased in underdeveloped nations and increases in the elderly population. It can be effectively treated with proton pump inhibitors and antibiotics.

328. The answer is e. *(Fauci, 14/e, pp 1619–1620.)* The presence of steatorrhea is diagnostic for malabsorption particularly when more than 6 g of fat is excreted per day. Villous atrophy on small bowel biopsy specimens is also consistent with changes leading to malabsorption. Calcium and iron are not absorbed well in the upper small intestine, and their levels are low. Similarly, the zinc level should be low as well.

329. The answer is e. *(Fauci, 14/e, p 1621.)* Causes of malabsorption include the absence of digestive enzymes (as seen in chronic pancreatitis) or injured or absent small bowel mucosa (as seen in Crohn's disease, various gastric surgeries, and ischemia resulting in bowel resection). Sigmoid resection or removal of all or part of the colon should not impair absorption of nutrients.

330. The answer is d. *(Fauci, 14/e, pp 1637–1639.)* Ulcerative colitis usually presents with diarrhea, frequently bloody. After 10 years of total colonic involvement, there is an increased risk of colon malignancy. Presentation usually is in a continuous fashion without skip lesions. Granulomas are not seen in biopsy specimens.

331. The answer is e. *(Fauci, 14/e, pp 1637–1639.)* Crohn's disease can present with disease confined to the small bowel or colon or more commonly with a combination of ileocolonic involvement. It frequently has skip lesions throughout the small bowel and colon, with abnormal mucosa separated by normal mucosa. It can present with nausea and vomiting secondary to obstruction due to fibrosing disease, stricture, or abscess.

332–335. The answers are a, b, c, d. *(Fauci, 14/e, pp 253–255.)* An ultrasound is the ideal test to examine the gallbladder for stones. CT scan is very helpful in evaluating obstructive jaundice, particularly when a pancreatic source is suspected. For subtle lesions for which the differential diagnosis includes metastatic disease, vascular problems, or parenchymal disease, MRI is very helpful if the CAT scan is not diagnostic. Barium swallow is the initial study favored in the evaluations of patients with dysphagia.

336. The answer is d. *(Fauci, 14/e, p 909.)* The presence of diarrhea and positive white cells are nonspecific and not diagnostic for *C. difficile*, which usually is seen with prior antibiotic usage, but this is not totally reliable. Identification of *C. difficile* toxins in the stools as well as the presence of pseudomembranes on sigmoidoscopy are pathognomonic.

337. The answer is e. *(Fauci, 14/e, p 1684.)* Patients with acute hepatitis A do not progress to chronic disease, unlike those with hepatitis B and C. There is a less than 1 percent mortality rate with acute hepatitis A, and there is no chronic disease secondary to this entity.

338. The answer is d. *(Fauci, 14/e, p 2151.)* Diagnostic iron studies for hemochromatosis include those for elevated ferritin, slightly elevated TIBC, and higher Fe levels resulting in a high saturation. Elevated ferritin usually over 500 is seen.

339. The answer is e. *(Fauci, 14/e, p 1708.)* Positive AMA is seen in approximately 95 percent of PBC patients. ANA is seen in a minority. Decreased ceruloplasmin and increased ferritin levels are usually seen in Wilson's disease and hemochromatosis, respectively. There is no association between hepatitis B surface antigen and PBC.

340. The answer is e. *(Fauci, 14/e, p 1699.)* The most serious consequences of chronic hepatitis B and C disease are hepatoma and cirrhosis. Complications of cirrhosis may occur, resulting in liver failure and inability to be gainfully employed. Hepatic adenoma is not a risk factor for these chronic diseases.

341. The answer is e. *(Fauci, 14/e, p 1677.)* Acute viral hepatitis, no matter what the etiology, usually presents in a similar fashion with fatigue and anorexia, malaise, and dark urine, if jaundice is present. Transaminases are usually 10 times above normal. Spider angiomas are seen as a late manifestation of chronic liver disease.

342. The answer is e. *(Fauci, 14/e, p 1691.)* Protective vaccines currently are available for hepatitis A and hepatitis B. Work is ongoing for hepatitis C vaccine. No such vaccines exist for hepatitis D or E.

343. The answer is e. *(Fauci, 14/e, p 1705.)* Acute alcoholic hepatitis is characterized by the presence of Mallory bodies and WBCs in the liver specimen. Frequently, there is fat as well as degenerating cells in the liver. Usually, with abstinence from alcohol, this will revert to a normal histology although, on occasion, liver disease can progress despite abstinence.

LIVER DISEASE

Questions

DIRECTIONS: Each item below contains a question or incomplete statement followed by suggested responses. Select the **one best** response, the **matching** response, or the **true/false** response to each question.

Items 344–348

Serum alkaline phosphatase may be elevated in the diseases of the following organs (answer True or False for each):

344. Bone

345. Intestine

346. Liver

347. Salivary glands

348. Spleen

349. Transmission of hepatitis A is almost exclusively by
a. blood transfusion
b. IV drug abuse
c. fecal–oral route
d. sexual

Items 350–353

Answer True or False for each of these statements about hepatitis B surface antigen (HBsAg):

350. It is the first serologic marker detectable in the serum after an infection

351. Elevation of serum transaminase activity and clinical symptoms precede the appearance of hepatitis B surface antigen in the serum.

352. It is an envelope protein.

353. In typical cases, HBsAg becomes undetectable 1 to 2 months following the onset of jaundice.

354. The most common cause of fulminant hepatitis is hepatitis
a. A
b. B
c. C
d. E
e. G

Items 355–359

Answer True or False for each of these statements about acetaminophen-induced liver damage:

355. Fatal fulminant disease is usually associated with ingestion of 25 g or more of acetaminophen.

356. Blood levels of acetaminophen do not correlate with the severity of hepatic injury.

357. Hepatotoxicity is mediated by a toxic reactive metabolite formed from the parent compound by the cytochrome P-450 system.

358. Therapy with N-acetylcysteine should preferably begin within 8 h but may be given as late as 24 to 36 h after ingestion.

359. N-Acetylcysteine acts by increasing the renal excretion of acetaminophen.

360. Answer True or False for each of these statements about chronic hepatitis B:
a. It is more likely to occur if infection occurs in adults.
b. Seroconversion from HBeAg positive to HBeAg negative after 4 months of interferon α therapy is 40 percent.
c. Long-term therapy with steroids is also effective.
d. The likelihood of responding to interferon is greater in patients with high levels of hepatitis B virus (HBV) DNA.

Items 361–363

Answer True or False for each of these statements about hepatitis C:

361. It is a DNA virus.

362. Chronic hepatitis follows acute infection in 50 to 70 percent of the cases.

363. Sustained response to 6 months of interferon therapy is approximately 30 to 40 percent.

364. In patients with alcohol-induced liver disease, all of the following are indicators of poor prognosis EXCEPT
a. ascites
b. hepatic encephalopathy
c. marked hyperbilirubinemia
d. normal prothrombin time

Items 365–368

Answer True or False for each of these statements about variceal bleeding:

365. About half of all episodes of variceal hemorrhage cease without intervention.

366. Somatostatin is not as effective as vasopressin in treating variceal bleeding.

367. Endoscopic sclerotherapy controls acute bleeding in 90 percent of the cases.

368. Emergency portal systemic shunt surgery has a low mortality of less than 20 percent in patients with variceal bleeding.

369. A serum–ascitic fluid albumin gradient of more than 1.1 g/dL is consistent with ascites caused by
a. tuberculosis
b. peritoneal metastases
c. cirrhosis of the liver
d. trauma

Items 370–374

Answer True or False for each of these statements about spontaneous bacterial peritonitis (SBP):

370. The ascitic fluid typically has a high concentration of albumin.

371. The ascitic fluid polymorphonuclear leukocyte count is greater than 250/µL.

372. Empirical therapy with cefotaxime should be initiated when the diagnosis is first suspected.

373. Infection is most commonly caused by pneumococci.

374. As many as 70 percent of the patients will experience at least one recurrence within a year of the first episode.

Items 375–378

Answer True or False for each of these statements about hepatorenal syndrome:

375. The urinary sodium is less than 5 mmol/L.

376. It does not occur without a precipitating factor.

377. It is characterized by azotemia, hyponatremia, progressive oliguria, and hypotension.

378. Urinary sediment has a high concentration of RBC casts.

Items 379–383

Match the following hepatitis viruses with their predominant mode of transmission:
a. Oral–fecal
b. Percutaneus/parenteral

379. Hepatitis A

380. Hepatitis B

381. Hepatitis C

382. Hepatitis D

383. Hepatitis E

Items 384–388

Primary biliary cirrhosis (PBC) has the following characteristics (answer True or False for each):

384. A circulating IgG antimitochondrial antibody is detected in more than 90 percent of the patients.

385. Among patients with symptomatic disease, 90 percent are men between the ages of 50 and 70 years of age.

386. Pruritis and fatigue are common early symptoms.

387. Serum alkaline phosphatase is elevated twofold to fivefold.

388. Glucocorticoids are effective in treating PBC.

LIVER DISEASE

Answers

344–348. The answers are t, t, t, f, f. *(Fauci, 14/e, p 1664.)* Human serum contains several forms of alkaline phosphatase, a plasma membrane-derived enzyme of uncertain physiologic function that hydrolyzes synthetic phosphate esters at pH 9. These activities arise from the bone, intestine, liver, and placenta. In the absence of bone disease and pregnancy, elevated levels of alkaline phosphatase activity usually reflect impaired biliary tract function.

349. The answer is c. *(Fauci, 14/e, p 1684.)* Hepatitis A is transmitted almost exclusively by the fecal–oral route. Person-to-person spread of hepatitis A virus (HAV) is enhanced by poor personal hygiene and overcrowding, and large outbreaks as well as sporadic cases have been traced to contaminated food, water, milk, and shellfish. Intrafamily and institutional spread are also common.

350–353. The answers are t, f, t, t. *(Fauci, 14/e, pp 1679–1680.)* After infection with HBV, the first virologic marker detectable in the serum is HBsAg. Circulating HBsAg precedes elevations of serum transaminase activity and clinical symptoms and remains detectable during the entire icteric or symptomatic phases of acute hepatitis B and beyond. In typical cases, HBsAg becomes undetectable in 1 to 2 months following the onset of jaundice and rarely persists beyond 6 months.

354. The answer is b. *(Fauci, 14/e, p 1689.)* Hepatitis B accounts for more than 50 percent of fulminant hepatitis cases, a sizable proportion of which are associated with hepatitis D virus (HDV) infection.

355–359. The answers are t, f, t, t, f. *(Fauci, 14/e, p 1694.)* Acetaminophen has caused severe hepatic necrosis when ingested in large amounts in suicide attempts and accidentally by children. Fatal fulminant disease is usually (although not invariably) associated with ingestion of 25 g

118   Pathophysiology

or more. Therapy should begin within 8 h of ingestion but may be effective if given as late as 24 to 36 h after overdose. N-Acetylcysteine appears to act by providing a reservoir of sulfhydryl groups to bind to toxic metabolites or by stimulating synthesis and repletion of hepatic glutathione.

360. The answer is a. *(Fauci, 14/e, pp 1698–1699.)* The likelihood of chronicity after acute hepatitis B varies as a function of age. Infection at birth is associated with a 90 percent chance of chronic infection, whereas infection in young adulthood in immunocompromised persons is associated with an approximately 1 percent risk of chronicity. The likelihood of responding to interferon is greater in patients with moderate to low levels of HBV DNA and in patients with substantial elevations of aminotransferase activity. In patients with hepatitis B, long-term therapy with glucocorticoids is not only ineffective but also detrimental.

361–363. The answers are f, t, f. *(Fauci, 14/e, pp 1700–1701.)* Hepatitis C is an RNA virus. Chronic infection follows acute infection in 50 to 70 percent of cases. Six-month treatment with interferon is associated with a likelihood of biochemical response in approximately 50 percent of the patients. Clinical trials have demonstrated that, after 6 months of therapy, at least 50 percent of the responding patients experience a biochemical relapse; based on this relapse rate, the likelihood of a sustained response would be 25 percent.

364. The answer is d. *(Fauci, 14/e, p 1706.)* In patients with alcoholic hepatitis, the presence of marked hyperbilirubinemia, rising serum creatinine level, marked prolongation of prothrombin time, ascites, and encephalopathy are associated with a poor short-term prognosis.

365–368. The answers are t, f, t, f. *(Fauci, 14/e, pp 1711–1712.)* Variceal bleeding is a life-threatening emergency. About half of all episodes of variceal hemorrhage cease without intervention, although the risk of rebleeding is very high. Samatostatin has been shown to be as effective as vasopressin. Emergency portal systemic nonselective shunts may control acute hemorrhage, but such surgery is usually used only as a last resort because early operative mortality is greater than 30 percent.

369. The answer is c. *(Fauci, 14/e, p 1713.)* The serum–ascites albumin gradient provides a better classification than total protein count or other parameters. Ascites resulting from cirrhosis of the liver typically has a high serum–ascites albumin gradient (less than 1.1 g/dL), reflecting indirectly the abnormally high hydrostatic pressure gradient between the portal bed and the ascitic compartment.

370–374. The answers are f, t, t, f, t. *(Fauci, 14/e, pp 1714–1715.)* Patients with cirrhosis and ascites develop SBP without any obvious primary source of infection. The ascitic fluid in these patients typically has a low concentration of albumin. An ascitic fluid count of more than 250 polymorphonuclear cells is suggestive of SBP. Empirical therapy with cefotaxime should be initiated when the diagnosis is first suspected, because enteric gram-negative bacteria are found in the majority of cases. Recurrent episodes are relatively common; as many as 70 percent of the patients will experience at least one recurrence within 1 year of the first episode.

375–378. The answers are t, f, t, f. *(Fauci, 14/e, p 1715.)* Worsening azotemia, hyponatremia, progressive oliguria, and hypotension are the hallmarks of the hepatorenal syndrome, which may be precipitated by severe GI bleeding, sepsis, or overly vigorous attempts at diuresis. It also may occur without any obvious cause. Typically, the urine sodium concentration is less than 5 mmol/L. The urinary sediment is unremarkable.

379–383. The answers are a, b, b, b, a. *(Fauci, 14/e, p 1697s.)* Hepatitis A and E are transmitted via the oral–fecal route, whereas hepatitis B, C, and D are transmitted via the percutaneous/parenteral route.

384–388. The answers are t, f, t, t, f. *(Fauci, 14/e, pp 1707–1708.)* The cause of PBC remains unknown. A circulating IgG antimitochondrial antibody is detected in more than 90 percent of the cases. Among patients with symptomatic disease, 90 percent are women between the ages of 35 and 60 years. In the treatment of PBC, glucocorticoids are ineffective and may actually worsen the bone disease.

THYROID AND PITUITARY DISORDERS

Questions

DIRECTIONS: The group of questions in this section consists of lettered options followed by numbered items. For each numbered item, select the **one best** lettered option. Each lettered option may be used once, more than once, or not at all.

Items 389–393
a. Graves' disease
b. Jodbasedow phenomenon
c. Choriocarcinoma
d. Struma ovarii
e. Toxic multinodular goiter

389. Hyperthyroidism in a patient who moved from an iodine-deficient area to an iodine-sufficient area

390. Thyrotoxicosis and uniformly increased radioactive iodine uptake in the thyroid without thyrotropin receptor antibodies

391. Pretibial myxedema

392. Infiltration of orbital soft tissue and extraocular muscles with lymphocytes, mucopolysaccharides, and fluid

393. Thyrotoxicosis having a low uptake of iodine in the thyroid bed but uptake in the pelvis

Items 394–398
a. Iodine deficiency
b. Lithium
c. Hashimoto's thyroiditis
d. Propylthiouracil
e. Toxic multinodular goiter

394. The most common cause of spontaneous hypothyroidism in the United States

395. The most common cause of goiter in developing nations

396. Endemic goiter

397. Inhibits conversion of T_4 to T_3

398. High levels of thyroidal peroxidase antibody

Items 399–403

a. Hyperthyroidism
b. Nonthyroidal illness (sick euthyroidism)
c. Estrogen therapy
d. Subclinical hypothyroidism
e. Familial (euthyroid) dysalbuminemic hyperthyroxinemia

399. Elevated TSH, normal free T_4, and no recent illnesses

400. Normal TSH, normal T_4, and low T_3

401. Low TSH, high T_4, and high T_3

402. Normal TSH, high T_4, and high T_3

403. Low TSH and high T_3

Items 404–408

a. Thyroid lymphoma
b. Medullary thyroid carcinoma
c. Papillary thyroid carcinoma
d. Anaplastic thyroid carcinoma
e. Follicular thyroid carcinoma

404. Most common thyroid cancer

405. Life expectancy of less than 6 months from diagnosis

406. Rapidly enlarging thyroid mass in a patient with chronic autoimmune (Hashimoto's) thyroiditis

407. Psammoma bodies

408. Elevated plasma calcitonin

Items 409–411

a. Graves' disease
b. Subacute thyroiditis
c. Toxic multinodular goiter
d. Hashimoto's thyroiditis
e. Toxic adenoma

409. Markedly tender gland, low radioiodine uptake, and thyrotoxicosis

410. Thyrotoxicosis with a patchy pattern but normal amount of radioiodine uptake

411. Diffusely enlarged gland with uniform and increased radioiodine uptake

412. A 45-year-old man presents because of frontal bossing and an enlarged nose, tongue, and jaw. He has doughy palms and spadelike fingers. The best screening test to establish the diagnosis is
a. random growth hormone
b. insulin-like growth factor (IGF-1)
c. TSH
d. prolactin
e. fasting blood sugar

413. Based on physical findings, you suspect that a 48-year-old woman has acromegaly. The definitive diagnostic test for acromegaly is measurement of growth hormone in the following setting:
a. Random
b. Thyrotropin-releasing hormone (TRH) stimulation test
c. Insulin tolerance test
d. Oral glucose tolerance test
e. Luteinizing hormone-releasing hormone (LHRH) stimulation test

414. You confirm acromegaly in a 58-year-old woman, and an MRI of the pituitary shows a microadenoma. The best choice for treatment is
a. transsphenoidal surgery
b. medical therapy with somatostatin agonist
c. irradiation
d. medical therapy with bromocryptine
e. transfrontal surgery

415. Untreated acromegaly results in decreased life expectancy for the following reasons EXCEPT
a. cerebral vascular disease
b. congestive heart failure
c. respiratory disease
d. colon carcinoma
e. cervical arthropathy

416. A 30-year-old woman presents with a 6-month history of amenorrhea. Your initial evaluation should include measurement of
a. prolactin
b. estradiol
c. progesterone
d. testosterone
e. DHEA-S

417. A 28-year-old woman develops galactorrhea without amenorrhea. Your evaluation should include
a. estradiol
b. progesterone
c. prolactin
d. testosterone
e. DHEA-S

418. A 47-year-old man has had a headache and experienced impotence for the past 2 months. A likely hormonal profile would be
a. low testosterone, high LH, and low prolactin
b. low testosterone, low LH, and low prolactin
c. low testosterone, high LH, and high prolactin
d. normal testosterone, normal LH, and normal prolactin
e. low testosterone, low LH, and high prolactin

419. A 28-year-old woman has amenorrhea and galactorrhea, after beginning a new medication recently. The most likely medication is
a. haloperidol
b. lisinopril
c. fluoxetine
d. amitriptyline
e. buspirone

420. A 25-year-old woman has amenorrhea and galactorrhea. The results of her thyroid function tests are normal. Her prolactin level is 350 µg/L (n <20). The most likely cause for her hyperprolactinemia is
a. microadenoma
b. macroadenoma
c. antidepressant use
d. exercise induced
e. antihypertensive therapy

421. A 26-year-old woman has been amenorrheic for 2½ months. Your first choice for diagnostic evaluation is
a. hCG
b. LH
c. estradiol
d. prolactin
e. progesterone

422. A 40-year-old man has erectile dysfunction. He is noted to have hyperprolactinemia (prolactin of 400 µg/L). On MRI, a macroadenoma with superstellar extension is found. The best course of therapy for the patient is
a. medical therapy with bromocriptine
b. transsphenoidal surgery
c. transfrontal surgery
d. medical therapy with somatostatin agonist
e. thyroxine

423. A 35-year-old man has a prolactinoma and a history of severe peptic ulcer disease. There is a family history of pituitary tumors. The findings of what other diagnostic test at this time may be abnormal and potentially useful in diagnosis?
a. Fasting blood sugar
b. Serum calcium
c. Serum calcitonin
d. Urinary metanephrine
e. Serum ferritin

424. A 45-year-old man has decreased libido and decreased sexual function. A large pituitary tumor is found. His prolactin is 20 (<15). Testing of his pituitary–gonadal axis most likely will demonstrate
a. normal testosterone and low LH
b. high testosterone and normal LH
c. low testosterone and low LH
d. normal testosterone and normal LH
e. low testosterone and high LH

425. A 16-year-old boy presents without pubertal development and development of secondary sexual characteristics. He cannot smell (anosmia). The baseline testosterone and the LH response to LHRH most likely are
a. low testosterone and normal LHRH response
b. normal testosterone and normal LHRH response
c. high testosterone and normal LHRH response
d. low testosterone and no LHRH response
e. low testosterone and exaggerated LHRH response

426. A 58-year-old woman presents as an outpatient with lethargy, fatigue, and cold intolerance. Thyroid function testing reveals a free T_4 level of 0.5 (0.7 to 2.0) and a TSH of 0.1 (0.5 to 5). The next best diagnostic test is
a. thyroid scan and uptake
b. MRI of the pituitary
c. prolactin
d. thyroid autoantibodies
e. T_3

427. A 59-year-old man presents with heat intolerance and tremor. Thyroid function testing reveals a free T_4 level of 3.0 (0.7 to 2.0) and TSH level of 6.0 (0.5 to 5). The next best diagnostic test is
a. thyroid scan and uptake
b. MRI of the pituitary
c. prolactin
d. thyroid autoantibodies
e. T_3

428. A 25-year-old woman presents with increasing obesity, amenorrhea, hypertension, and abdominal stria. The next best diagnostic test is
a. prolactin
b. free T_4 and TSH
c. overnight dexamethasone suppression
d. random cortisol
e. adrenocorticotropic hormone (ACTH)

429. A 30-year-old man presents with weight gain, dorsocervical fat pad, and proximal muscle weakness. His urinary free cortisol level is markedly elevated and does not suppress with dexamethasone. The plasma ACTH is undetectable. Your best next diagnostic test is
a. serum antidiuretic hormone (ADH)
b. chest CT
c. MRI of the pituitary
d. ACTH stimulation test
e. abdominal CT

430. A 65-year-old man with a lung mass has increasing skin pigmentation and marked muscle weakness and wasting. His urinary free cortisol level is 690 µg/24 h (10 to 80) and is nonsuppressible. Which of the following laboratory tests would probably be most diagnostic?
a. ACTH stimulation test
b. MRI of the pituitary
c. CT of the abdomen
d. plasma ACTH
e. parathyroid hormone

431. A 48-year-old woman with a history of pituitary surgery and irradiation is scheduled for elective surgery. She currently requires replacement thyroxine, hydrocortisone, estrogen, and progesterone. In the perioperative period, you will treat her with
a. glucose infusion
b. increased hydrocortisone
c. ACTH infusion
d. increased estrogen
e. increased thyroxine

432. A 23-year-old woman presents with weakness and amenorrhea. She is clinically hypothyroid. A CT scan of the pituitary shows an expanded sella with a large cystic component with calcifications. The most likely diagnosis is
a. pituitary macroadenoma
b. empty-sella syndrome
c. craniopharyngioma
d. optic glioma
e. hypothalamic hamartoma

433. Patients with pituitary macroadenoma present most commonly with
a. bitemporal hemianopsia
b. unilateral optic atrophy
c. left or right homonymous visual field defect
d. unilateral center scotoma
e. left or right superior temporal defect

434. A 45-year-old man has decreased libido and erectile dysfunction. He has noted increasing pigmentation. He has developed liver disease and arthropathy recently. The next best diagnostic test is
a. serum TSH
b. serum calcium
c. serum prolactin
d. serum ferritin
e. serum gastrin

THYROID AND PITUITARY DISORDERS

Answers

389–393. The answers are b, c, a, a, d. (*McPhee, 2/e, pp 475–479.*) Thyrotoxicosis can have several etiologies. Iodine-induced hyperthyroidism is called the Jodbasedow phenomenon and can occur in patients with endemic goiter who move to areas where iodine is plentiful. Diffusely increased radioiodine uptake in the thyroid accompanying thyrotoxicosis usually indicates Graves' disease, in which the thyrotropin receptors are stimulated by antibodies. In patients with choriocarcinoma, however, high levels of human chorionic gonadotropin (hCG) can also stimulate the thyrotropin receptor and produce the same finding. Graves' disease is associated with related autoimmune phenomena in other tissues, such as Graves' ophthalmopathy in the orbit and pretibial myxedema in the skin. Ovarian teratomas can contain thyroid tissue, struma ovarii, and rarely cause thyrotoxicosis, with excess thyroid hormone produced by the teratoma rather than the thyroid.

394–398. The answers are c, a, a, d, c. (*McPhee, 2/e, pp 480–486; Fauci, 14/e, pp 2021–2023.*) Hypothyroidism can result from several causes, including congenital defects, chronic autoimmune thyroiditis (Hashimoto's thyroiditis), medications (thionamides, lithium, or iodine), other iatrogenic causes, iodine deficiency, and hypothalamic or pituitary insufficiency. Frequently, goiter is associated with hypothyroidism, but goiter can also be found in euthyroid or hyperthyroid patients. Worldwide, iodine deficiency (endemic) goiter is very common. Chronic autoimmune thyroiditis, which is the most common cause of hypothyroidism in the United States, is associated with high levels of thyroid autoantibodies. Propylthiouracil propranolol, glucocorticoids, and iodine inhibit conversion of T_4 to T_3.

399–403. The answers are d, b, a, c, a. (*McPhee, 2/e, pp 470–475, 485–486; Fauci, 14/e, pp 2016–2019.*) Laboratory measurements of thyroid hormones and TSH have proven invaluable in determining the true functional status of the thyroid gland. However, various medications and nonthyroidal illnesses can alter certain values, so usually a combination of values is used to make a diagnosis. TSH values tend to be the most reliable in the absence of hypothalamic or pituitary disease, and mild elevation is seen in hypothyroidism before free T_4 declines. In severe nonthyroidal illness, T_3 declines first, followed by T_4 if the disease is severe enough, but TSH is usually normal. Low TSH with high T_4 and T_3 or T_3 alone (T_3 toxicosis) reflects hyperthyroidism. Estrogens increase thyroxine-binding globulin, elevating total T_4 and T_3, while free T_4, free T_3, and TSH remain normal.

404–408. The answers are c, d, a, c, b. *(Fauci, 14/e, pp 2030–2033.)* Thyroid cancers may arise from the thyroid follicular epithelium, the parafollicular C cells, or lymphoid cells in the thyroid. Papillary carcinomas, including tumors with mixed papillary and follicular elements, are most common and account for 70 percent of thyroid cancers. Psammoma bodies are a feature of papillary carcinomas. Of thyroid cancers, 15 percent have a purely follicular histology. Medullary thyroid carcinomas arise in the calcitonin-producing parafollicular cells and account for about 5 percent of thyroid cancers. Thyroid lymphomas constitute about 5 percent of thyroid cancers and occur most often in patients with Hashimoto's thyroiditis. Lymphomas and anaplastic carcinomas tend to grow rapidly. The prognosis of anaplastic cancers, which likely represent dedifferentiation of better differentiated papillary or follicular carcinomas, is very poor, with average survival less than 6 months.

409–411. The answers are b, c, a. *(Fauci, 14/e, pp 2016, 2023–2030, and 2033–2034.)* The pattern and amount of radioiodine uptake on ^{123}I scan is fundamental to the correct diagnosis of thyrotoxicosis. Low-uptake thyrotoxicosis can occur when there is destruction of the thyroid follicles with release of thyroid hormone, such as in subacute thyroiditis, which usually presents as an exquisitely painful gland. Iodine-induced hyperthyroidism, factitious hyperthyroidism, and painless (silent) thyroiditis also cause low-uptake thyrotoxicosis. Patchy radioiodine uptake is common in multinodular goiter and Hashimoto's thyroiditis, but hyperthyroidism with normal or increased uptake typifies toxic multinodular goiter. In Graves' disease, the uptake tends to be increased and more uniform. Uptake may be increased without thyrotoxicosis in conditions characterized by defects in organification of iodine, such as is found in some patients with Hashimoto's thyroiditis.

412. The answer is b. *(Fauci, 14/e, p 1982.)* The best screening test for suspected acromegaly is a test for IGF-1. Random growth hormone varies too much to be useful. TSH and prolactin may be abnormal but are not

130 Pathophysiology

diagnostic of acromegaly. Fasting blood sugar may be elevated in this patient, but again it is not diagnostic.

413. The answer is d. *(Fauci, 14/e, p 1982.)* The most definitive and widely accepted test for the diagnosis of acromegaly is the response of growth hormone during an oral glucose tolerance test. Typically, the growth hormone at baseline in acromegaly will be above 5 µg/L. In normal patients, the growth hormone will suppress to less than 2. In acromegalic patients, the growth hormone values may rise, show no change, or suppress partially but not less than 2. A single random growth hormone is not useful, because of the pulsatility in growth hormone. TRH does stimulate growth hormone in many acromegalics but not all. The insulin tolerance test is a stimulation test of growth hormone and not a suppression test.

414. The answer is a. *(Fauci, 14/e, p 1982.)* Transsphenoidal surgery has the advantages of potential cure with rapid therapeutic response. If the tumor is completely resected, the patient may experience a complete cure. Medical therapy with somatostatin agonist or bromocryptine is helpful, but the patient is dependent on medical therapy indefinitely. Irradiation takes years for full effectiveness, and the patient may develop hypopituitarism. Transfrontal surgery is rarely employed now.

415. The answer is e. *(Fauci, 14/e, p 1981.)* Patients with untreated acromegaly have shortened life expectancy and develop complications of cardiovascular, cerebrovascular, and respiratory disease. There are recent studies suggesting patients with acromegaly have increased frequency of polyps and subsequent development of colon carcinoma. Bowel surveillance has been suggested. Cervical arthropathy is a frequent complication of acromegaly but does not directly decrease life expectancy.

416. The answer is a. *(Fauci, 14/e, p 1975.)* A common presentation for hyperprolactinemia is amenorrhea. Important in the initial evaluation of amenorrhea is a determination of the prolactin level. Estradiol and progesterone typically are not measured in an initial evaluation of amenorrhea. Testosterone and DHEA-S are markers for androgen excess, which may be present in this patient, but do not need to be measured initially.

417. The answer is c. *(Fauci, 14/e, p 1975.)* Galactorrhea in young females is often associated with hyperprolactinemia. Estradiol and progesterone can be useful markers of gonadal function but not provide further diagnostic information. Similarly, testosterone and DHEA-S do not provide more diagnostic information.

418. The answer is e. *(Fauci, 14/e, p 1975.)* Men frequently present with marked hyperprolactinemia from a macroadenoma. Presenting manifestations typically are sexual dysfunction and decreased libido. The prolactin causes a decrease in LH and a concomitant decrease in testosterone. Thus, the patient will have a high prolactin level associated with low LH and low testosterone levels. The pattern of low testosterone, high LH, and low prolactin is typical of primary hypergonadism.

419. The answer is a. *(Fauci, 14/e, p 1975.)* Medications are important in the differential diagnosis of hyperprolactinemia. A common drug that causes increased prolactin with possible amenorrhea and galactorrhea is haloperidol, a dopamine antagonist. Lisinopril has no effect on prolactin levels. The antidepressants fluoxetin and amitriptyline and anxiolytic buspirone may cause small changes in prolactin levels but rarely enough to cause a clinical syndrome.

420. The answer is b. *(Fauci, 14/e, p 1975.)* The serum level of prolactin correlates roughly with the size of the tumor. Prolactin levels greater than 300 µg/L are most likely associated with macroadenoma. Increases in prolactin due to medications are usually less than 100. Microadenomas usually do not exceed levels of 200 to 300.

421. The answer is a. *(Fauci, 14/e, p 1975.)* The first choice in testing in this patient is a pregnancy test. If her prolactin level was measured without a pregnancy test, an elevation of prolactin could be wrongly considered primary rather than due to pregnancy. The other tests of LH, estradiol, and progesterone are not first choices in the evaluation of amenorrhea.

422. The answer is a. *(Fauci, 14/e, p 1976.)* Established therapy of hyperprolactinemia from a pituitary adenoma is treatment with a dopamine agonist, such as bromocriptine. Surgical therapy usually does not result in a cure in a macroadenoma and is reserved for those patients who are intolerant to dopamine agonist. Transfrontal surgery is rarely utilized. Somatostatin agonist and thyroxine have little effect on hyperprolactinemia.

423. The answer is b. (*Fauci, 14/e, pp 1975 and 2131.*) This patient may have multiple endocrine neoplasia syndrome-1, which presents with pituitary tumors, pancreatic tumors, and hyperparathyroidism. With the history of severe peptic ulcer disease (possible Zollinger-Ellison syndrome) and family history of pituitary tumors, one must suspect MEN-1. A serum calcium determination will be useful in diagnosing potential hyperparathyroidism. Elevated calcitonin and urinary metanephrine levels are characteristic of MEN-2. Serum ferritin and fasting blood sugar levels would be elevated in hemochromatosis.

424. The answer is c. (*Fauci, 14/e, p 1984.*) This patient has a common presentation for secondary hypogonadism. The large tumor is inhibiting LH secretion, with consequently low testosterone secretion. No other pattern fits this clinical presentation.

425. The answer is a. (*Fauci, 14/e, p 1984.*) This patient most typically has Kallmann's syndrome, which is a deficiency in the secretion of LHRH from the hypothalamus. Typically, these patients will respond to LHRH, although they may need LHRH priming. Testosterone will be low from the lack of LHRH stimulation of LH secretion.

426. The answer is b. (*Fauci, 14/e, p 1985.*) This patient presents with clinical manifestations of hypothyroidism, with a low free T_4. Secondary hypothyroidism is suggested by the low TSH. The diagnostic test of choice is an MRI of the pituitary for evaluation of a possible pituitary tumor.

427. The answer is b. (*Fauci, 14/e, p 1985.*) This patient likely has pituitary TSH-induced hyperthyroidism. This is a rare diagnosis, but the pattern of elevated free T_4 and elevated TSH is nearly diagnostic for this disorder. The next diagnostic test is an MRI of the pituitary to evaluate for the presence of microadenoma or macroadenoma.

428. The answer is c. (*Fauci, 14/e, p 1986.*) This patient presents with a high suspicion for Cushing's syndrome. The initial step in the evaluation should be an overnight dexamethasone suppression test. Failure to suppress would indicate a high likelihood of Cushing's syndrome. A random cortisol test is not sufficient to screen for Cushing's syndrome. An ACTH test by itself is not useful.

429. The answer is e. (*Fauci, 14/e, p 1986.*) The suppression of ACTH is characteristic of adrenal adenoma or carcinoma. A CT scan will evaluate for the presence of adrenal tumor. Chest CT is useful in determining ectopic ACTH secretion. An MRI of the pituitary is useful in pituitary-dependent Cushing's disease. The ACTH stimulation test and serum ADH evaluation are not diagnostic in this disease.

430. The answer is d. (*Fauci, 14/e, p 1986.*) This patient's presentation suggests ectopic ACTH secretion, and an ACTH will likely be elevated above 300. An MRI of the pituitary and CT of the abdomen are not useful, since the source of ACTH is from the small cell carcinoma in the lung mass.

431. The answer is b. (*Fauci, 14/e, p 1994.*) This patient has a classic history for hypopituitarism. During surgical stress, she will require an increased replacement dose of steroids. The other treatments will not cover her need for increased glucocorticoids and will not be helpful.

432. The answer is c. (*Fauci, 14/e, p 1988.*) This is the common CT finding and clinical presentation for craniopharyngioma. Empty sella does not usually cause marked enlargement of the sella, and there is no cystic structure with calcification. Pituitary macroadenoma can expand the sella but is not commonly cystic and calcified. Optic glioma and hypothalamic hamartoma are rarely cystic.

433. The answer is a. (*Fauci, 14/e, p 1990.*) The classic presentation is bitemporal hemianopsia, with the other visual field disturbances less common.

434. The answer is d. (*Fauci, 14/e, pp 1993 and 2150.*) This patient has classic manifestations of hemochromatosis, which impairs hypothalamic pituitary function. Evaluation of serum ferritin is potentially diagnostic in this patient. All the other tests are not diagnostic for hemochromatosis.

FEMALE AND MALE REPRODUCTIVE TRACTS

Questions

DIRECTIONS: Each item below contains a question or incomplete statement followed by suggested responses. Select the **one best** response or the **matching** response to each question.

435. Which of the following organs is not a major estrogen-dependent tissue in women?
a. Brain
b. Thyroid
c. Hypothalamus
d. Pituitary
e. Ovaries

436. Which of the following organs does not require androgens for proper growth in males?
a. Brain
b. Prostate
c. Epididymis
d. Vas deferens
e. Long bones

437. A patient has an excess of 17α-hydroxyprogesterone and 17α-hydroxypregnenolone in the urine, and no androgens. Which enzyme is deficient?
a. 20,22-Desmolase
b. 3β-Hydroxysteroid dehydrogenase
c. 17-Hydroxylase
d. 17,20-Desmolase (17,20-lyase)
e. 17-Ketoreductase

438. Which of these hormones is also produced in significant amounts outside of the gonads?
a. Estrone
b. Estradiol
c. Androstenedione
d. Testosterone
e. Dihydrotestosterone

439. A 22-year-old female marathon runner has had amenorrhea for 8 months. There has been no weight change, and her serum pregnancy test is negative. She has never been pregnant. Her menarche was at 13 years of age, and she had monthly menses until 8 months ago. Physical exam shows a women who is 66 inches tall, weighs 90 pounds, and is otherwise fully normal. Why does she have amenorrhea?
a. Hypothyroidism
b. Prolactinoma
c. Early menopause
d. Resistance to LH and follicle-stimulating hormone (FSH)
e. Excessive exercise

135

440. A young couple, both in their 20s, have been trying for 2 years to have a baby. The man comes into the office and, on workup, has oligospermia, a high LH level, a high FSH level, and a normal karyotype. How do you treat him?
a. Do nothing.
b. Administer testosterone injections.
c. Check the partner for causes of infertility.
d. Counsel regarding infertility.

441. A 28-year-old woman has had 2 days of abdominal pain and a positive pregnancy test. Her last menstrual period was 9 weeks ago. She reports no dysuria. She reports a history of two episodes of pelvic inflammatory disease. Which of these is the most likely cause of the abdominal pain?
a. Endometriosis
b. Urinary tract infection
c. Ectopic pregnancy
d. Placental abruption
e. Premenstrual syndrome

442. A 38-year-old woman has had amenorrhea for 6 months, with increased cold intolerance, loss of energy, and hair loss. Her menses were normal until this episode started, and she has also gained 22 pounds over these 6 months. Her pregnancy test is negative. Which test would you now order?
a. FSH and LH levels
b. Estrogen level
c. Testosterone level
d. TSH level
e. Cortisol level

443. A 29-year-old women has had three spontaneous (unplanned) abortions. All three occurred at approximately 6 weeks gestational age. The results of her physical exam are normal. Which of these may be the cause?
a. Ovary
b. Thyroid gland
c. Adrenal gland
d. Pituitary gland

444. Which of these medications treats benign prostatic hyperplasia by 5α-reductase inhibition?
a. Leuprolide
b. Nafarelin
c. Flutamide
d. Megesterol
e. Finasteride

445. How does pregnancy increase the risk of diabetes mellitus?
a. Causes weight gain
b. Insulin resistance
c. Placental production of human chorionic somatomammotropin
d. Increase of maternal glucocorticoids

446. Which of these hormones blocks milk production during pregnancy?
a. Progesterone
b. Prolactin
c. Chorionic somatomammotropin
d. Thyroxine
e. Insulin

447. A 40-year-old woman has amenorrhea and hirsuitism. Which hormone is in excess in this woman with polycystic ovary syndrome?
a. Estrogen
b. Progesterone
c. FSH
d. Androgens

448. Which of these is the best therapy for preeclampsia?
a. Antihypertensive medication
b. Antiseizure medication
c. IV fluids
d. Delivery of the baby
e. Anticoagulation

449. Which of these findings is the best for diagnosis of a congenital absence of the vas deferens?
a. Oligospermia
b. Azospermia
c. Normal testosterone
d. Normal LH
e. Normal FSH

450. Which of the following causes of male infertility is not caused by androgen deficiency?
a. Loss of libido
b. Inability to ejaculate
c. Azospermia
d. Congenital absence of the vas deferens
e. Sperm antibodies

451. A 19-year-old woman has galactorrhea. She has never been pregnant. Which hormone is the most likely to be responsible for this situation?
a. Prolactin
b. Estrogen
c. Progesterone
d. Thyroxine
e. Cortisol

138 Pathophysiology

452. A 15-year-old boy has a lack of pubic hair growth. He also informs you that his voice has not yet deepened, and he has no interest in sexual activity. He is an only child. Blood drawn reveals a very high testosterone level. What is the problem?
a. Low FSH and LH
b. High FSH and LH
c. Androgen insensitivity
d. Hyperthyroidism
e. XXY karyotype

453. A 52-year-old woman is having hot flashes. You suspect menopause. Which of the findings below would confirm your diagnosis?
a. Normal androgen level
b. Normal or low estrogen level
c. Normal prolactin level
d. High FSH and LH levels
e. High androgen level

454. Which of the following organs does ovarian estrogen production have a stimulatory effect on?
a. Ovary
b. Brain
c. Hypothalamus
d. Pituitary
e. Vagina

455. A 19-year-old pregnant woman has a blood clot in her leg. Which of the following coagulation factors is not increased in pregnancy?
a. Factor VII
b. Factor VIII
c. Factor IX
d. Factor X
e. Factor XI

456. Which of the following hormones is produced by both the ovary and uterus?
a. Inhibin
b. Activin
c. Follistatin
d. Relaxin
e. Enkephalin

457. Which of the following hormones is produced both by the theca and the kidney?
a. Inhibin
b. Relaxin
c. Renin
d. Epidermal growth factor-like
e. Transforming growth factor β

458. A 31-year-old man is infertile. His medical history reveals that he has Kartagener's syndrome. Why is he infertile?
a. Oligospermia
b. Asthenospermia
c. Absence of the vas deferens
d. Epididymal obstruction
e. Undescended testes

459. A 74-year-old man has had trouble urinating for 1 week. The force of the urinary stream is reduced, but there is no difficulty starting the stream. There is no pain. What is the problem?
a. Decreased detrusor contractility
b. Detrusor instability
c. Detrusor failure
d. Acute urinary obstruction
e. Chronic urinary obstruction

460. A 65-year-old man has benign prostatic hypertrophy and new-onset hypertension. Which medication could you give this patient to handle both diagnoses?
a. Nafarelin
b. Flutamide
c. Finasteride
d. Megesterol
e. Prazosin

461. A 28-year-old woman believes that she might be infertile. She had a healthy child 3 years ago and has been trying to get pregnant with the child's father for the last 18 months. There is no dysmenorrhea. Her menses occur regularly but have significantly less flow as compared with prior to the pregnancy. She recalls having a curettage performed to remove placental remnants. What is the diagnosis?
a. Ovarian failure
b. Hypothyroidism
c. Asherman's syndrome
d. Endometriosis
e. Prolactinoma

462. Which of the following is not part of the HELLP syndrome?
a. Hemolysis
b. Hypertension
c. Elevated liver enzymes
d. Thrombocytopenia

463. Which of the following is not stimulated by LH?
a. Thecal cells of the ovary
b. Granulosa cells of the ovary
c. Ovulation
d. Corpus luteum
e. Leydig cells of the testicle

464. Which of the following is not stimulated by FSH?
a. Ovarian follicles
b. Granulosa cells of the ovary
c. Ovulation
d. Leydig cells of the testicle
e. Sertoli cells of the testicle

465. A 28-year-old previously healthy woman, with no past medical history, is now 28 weeks pregnant. She complains of trouble seeing, polyuria, polyphagia, and polydipsia. What is her diagnosis?
a. Gestational diabetes mellitus
b. Deep venous thrombosis
c. Urinary tract infection
d. Preeclampsia

466. A 26-year-old woman is about to deliver a baby. She asks you why should she breast feed. Beyond the obvious issue of child–mother bonding, what else must you tell her?
a. It is the correct thing to do.
b. It provides better nutrition for the baby.
c. It protects the baby from infections early in life.
d. It will result in a quicker weight loss for the mother.
e. It is good contraception.

467. Which of the following is not required for successful breast feeding?
a. Prolactin
b. Oxytocin
c. Good maternal nutrition
d. An intact neurologic axis
e. High estrogen levels

468. Which of the following hormones in excess causes male infertility?
a. FSH
b. LH
c. Testosterone
d. Cortisol
e. Prolactin

469. How do varicoceles cause male infertility?
a. By decreasing testicular blood flow
b. By increasing testicular temperature
c. By reducing testosterone production
d. By causing testicular atrophy

470. An 18-year-old woman is seeking birth control pills. Which of the following is not an effect of these medications?
a. Nausea
b. Blood clots
c. Increased risk of cancer
d. Changes in the cervical mucus
e. Protection from sexually transmitted diseases

471. A 52-year-old woman has had a chief complaint of postmenopausal bleeding for 2 weeks. Her menopause was 7 years ago and, until 2 weeks ago, had no vaginal bleeding of any kind. She was not on hormonal replacement. Which of the following tests need not be ordered or performed?
a. Pap smear
b. TSH evaluation
c. Prothrombin time
d. Smear for vaginal infection
e. General blood chemistry evaluation

Items 472–476

Match each of the ovarian compartments with the hormone that it produces:
a. Granulosa
b. Theca
c. Both
d. Neither

472. Müllerian-inhibiting substance

473. Plasminogen activator

474. Renin

475. Basic fibroblast growth factor

476. Inhibin

Items 477–481

Match each of the causes of male infertility with the general cause:
a. Pretesticular
b. Testicular
c. Posttesticular

477. Low LH

478. Alcohol

479. Mumps

480. Phenytoin

481. Hypospadias

FEMALE AND MALE REPRODUCTIVE TRACTS

Answers

435. The answer is b. *(McPhee, 2/e, p 519.)* The thyroid gland is minimally affected by estrogens, but may have an impact on estrogen production (still unclear). The brain, hypothalamus, pituitary, ovaries, uterine epithelium, uterine tubes, and vagina are all major estrogen-dependent tissues in women.

436. The answer is a. *(McPhee, 2/e, p 547.)* The brain's growth is not affected by androgens, although there are mental changes associated with androgens. The prostate, epididymis, vas deferens, scrotum, seminal vesicles, penis, and long bones all require androgens for proper growth and physical development.

437. The answer is d. *(McPhee, 2/e, p 545; Fauci, 14/e, pp 2036 and 2100.)* 20,22-Desmolase changes cholesterol to pregnenolone. 3β-Hydroxysteroid dehydrogenase changes pregnenolone to progesterone and 17α-hydroxypregnenolone into 17α-hydroxyprogesterone. 17-Hydroxylase changes pregnenolone and progesterone into 17α-hydroxypregnenolone and 17α-hydroxyprogesterone respectively. 17,20-Desmolase changes 17α-hydroxypregnenolone into dehydroepiandrosterone (a weak androgen) and 17α-hydroxyprogesterone into androstenedione (a weak androgen). 17-Ketoreductase changes dehydroepiandrosterone and androstenedione into androstenediol and testosterone, respectively.

438. The answer is c. *(McPhee, 2/e, pp 491, 520, and 545; Fauci, 14/e, p 2036.)* In women, estrone and estradiol are produced mainly in the ovaries. Testosterone and dihydrotestosterone are mainly produced in the testes. Dihydrotestosterone is also produced in the periphery directly from testosterone. Androstenedione is the main end product of the zona reticularis of the adrenal gland, in addition to being produced in the gonads.

439. The answer is e. *(McPhee, 2/e, pp 531–535; Fauci, 14/e, pp 2105–2107.)* Hypothyroid patients tend to gain weight. Prolactin-secreting tumors (prolactinomas), being located in the pituitary, would be expected to show abnormal physical exam findings at the eyes, given that the tumor typically sits on the optic chiasma. Early menopause is unlikely in a 22-year-old. Resistance to LH and FSH would have prohibited this patient from ever having menses. This leaves excessive exercise as the only remaining plausible cause in this patient.

440. The answer is b. *(McPhee, 2/e, p 552; Fauci, 14/e, pp 2092–2097.)* This patient has classic testosterone deficiency, as evidenced by the low sperm count and elevated gonadotropic hormones.

441. The answer is c. *(McPhee, 2/e, pp 530 and 539; Fauci 14/e, pp 812–817.)* Pelvic inflammatory disease is a cause of tubal scarring, setting the stage for an ectopic (tubal) pregnancy. As the pregnancy grows, the tube is stretched, causing pain. Endometriosis causes pain with each menstrual cycle, which is not the case here. Urinary tract infection can cause pain, but she would be expected to have dysuria. Placental abruption occurs only after 20 weeks of pregnancy. She is clearly not premenstrual, because she is pregnant.

442. The answer is d. *(McPhee, 2/e, pp 481 and 532–534; Fauci 14/e, pp 2021–2023.)* Hypothyroidism is the cause of this patient's amenorrhea. Classic findings of hypothyroidism presented here are increased cold intolerance, loss of energy, and hair and weight gain. The best test for this disorder is that for TSH.

443. The answer is a. *(McPhee, 2/e, p 523.)* Once the woman is pregnant, she needs to maintain a high level of progesterone in her system to sustain the fetus. The corpus luteum, sitting in the ovary, has that role, under the influence of the β-hCG produced by the placenta. If the corpus luteum can not produce enough progesterone to get the pregnancy to week 10, the pregnancy is lost.

444. The answer is e. *(McPhee, 2/e, p 555; Fauci 14/e, pp 596–598.)* Leuprolide, nafarelin, and megesterol all inhibit LH secretion and thus decrease testosterone and dihydrotestosterone levels. Flutamide and megesterol both are androgen receptor inhibitors. Finasteride blocks 5α-reductase, leading to a reduction of dihydrotestosterone and net reduction of prostate size.

445. The answer is c. (*McPhee, 2/e, p 525; Barron, 2/e, pp 63–65.*) Human chorionic somatomammotropin is a counterregulatory hormone that works to protect the fetus from hypoglycemia. The net result can be hyperglycemia and thus diabetes mellitus. Weight gain does occur in pregnancy, but insulin resistance and increased production of maternal glucocorticoids have not been proven to occur.

446. The answer is a. (*McPhee, 2/e, pp 525–526; Fauci, 14/e, pp 2115–2116.*) All of these hormones are required for proper preparation of the breast to produce milk in the postpartum period, but high levels of progesterone and estrogen during pregnancy prevent actual milk production. Milk is produced post partum, once the levels of these two hormones drop.

447. The answer is d. (*McPhee, 2/e, pp 532–535; Fauci, 14/e, pp 2106–2107.*) The estrogen level is usually elevated from nonovarian sources. The progesterone level is usually unchanged. The FSH level is usually low. The androgen level is usually elevated and a cause for the symptoms here. These androgens are produced in the ovary as the result of an elevated LH level. A low FSH level prevents the formation of ovarian estrogen.

448. The answer is d. (*McPhee, 2/e, pp 539–540; Barron 2/e, pp 13–14.*) Although antihypertensive medication, antiseizure medication, intravenous fluids, and anticoagulation all may take care of parts of preeclampsia, the best therapy is to deliver the baby.

449. The answer is b. (*McPhee, 2/e, p 552; Fauci 14/e, pp 2092–2095.*) Whereas all of these may be present in patients with an absent vas deferens, azospermia is the best choice. Oligospermia, normal testosterone level, normal LH level, and normal FSH level may also be present in a wide variety of other causes of male infertility, alone or in combination.

450. The answer is e. (*McPhee, 2/e, pp 550–551.*) Libido, ejaculation, sperm production and fetal development of the male genital tract all require androgen. Sperm antibodies, be they from the male or the female, are immune mediated.

451. The answer is a. (*McPhee, 2/e, pp 461–462 and 525–526; Fauci 14/e, pp 2116–2117.*) Prolactin is the major stimulator of breast-milk production.

Overproduction of prolactin leads to galactorrhea. Estrogen, progesterone, thyroxine, and cortisol are all needed for proper breast development but play no role in actual milk production.

452. The answer is c. (*McPhee, 2/e, pp 544–548; Fauci, 14/e, pp 2088 and 2091–2092.*) Androgen insensitivity can present as an inability for a male child to go into puberty. Low FSH and LH levels are expected to yield low testosterone levels. High FSH and LH levels are usually markers of end-organ damage and lack of feedback of testosterone on the pituitary due to low testosterone levels. Puberty can be delayed by hyperthyroidism, with FSH and LH levels usually appropriate to the testosterone level. An XXY karyotype (Klinefelter's) often has no effect on testosterone level.

453. The answer is d. (*McPhee, 2/e, pp 527–528; Fauci, 14/e, p 2012.*) In both premenopausal and menopausal women, androgen levels may be normal or high, estrogen levels normal, and the prolactin level normal. Androgen production can be in the ovary or the adrenal gland. Estrogen production is mainly in the ovary prior to menopause and in the periphery by conversion of testosterone in menopause. The amount of estrogen during menopause is a function of the patient's amount of adipose tissue. High FSH and LH levels are the hallmarks of a lack of adequate ovarian production of both estrogen and progesterone, the chemical markers of menopause, due to lack of negative feedback.

454. The answer is e. (*McPhee, 2/e, pp 521–522.*) Estrogen has a stimulatory effect on the vagina, uterus, uterine tubes, and breasts. The effect on the ovary is paracrine. There is a negative feedback effect on the brain, hypothalamus, and pituitary.

455. The answer is e. (*McPhee, 2/e, p 529; Barron, 2/e, pp 241–244.*) Factors I, VII, VIII, IX, X, and XII are increased in pregnancy. Factors II, III, IV, V, VI, XI, and XIII are unchanged.

456. The answer is d. (*McPhee, 2/e, p 516.*) Relaxin is the only hormone produced both by the ovary and the uterus. Inhibin is produced in the granulosa, theca, and corpus luteum. Activin is produced in the granulosa. Follistatin is produced in follicles. Enkephalin is produced by the ovary.

457. The answer is c. *(McPhee, 2/e, pp 377 and 516.)* Renin is the only hormone produced by both the theca and the kidney. Inhibin is produced in the granulosa, theca, and corpus luteum. Relaxin is produced in the corpus luteum, theca, placenta and uterus. Epidermal growth factor-like is produced in the granulosa and theca. Transforming growth factor β is produced in the theca, ovarian interstitial tissue, and granulosa.

458. The answer is b. *(McPhee, 2/e, p 551; Fauci, 14/e, pp 1446 and 2092.)* Kartagener's syndrome is also known as the immotile cilia syndrome. Asthenospermia—poor sperm motility—is due to missing dynein arms, the basic defect of Kartagener's syndrome. Kartagener's syndrome has no effect on sperm count or the basic anatomy of the male reproductive tract.

459. The answer is a. *(McPhee, 2/e, pp 557–558; Fauci, 14/e, pp 262–265.)* Detrusor instability, decreased contractility, and failure are all part of a continuum. Decreased contractility is implied by the decreased force of the stream. Instability alone has only frequency and urgency. Failure implies an inability to urinate due to muscle failure. With acute obstruction, the patient cannot void, and there is significant pain. With chronic urinary obstruction, starting the stream is also a problem.

460. The answer is e. *(McPhee, 2/e, p 557; Fauci, 14/e, p 598.)* Alpha blockers, such as prazosin, can treat both hypertension and benign prostatic hypertrophy. Nafarelin, flutamide, finasteride, and megesterol have no role in blood pressure management.

461. The answer is c. *(McPhee, 2/e, pp 531–539; Fauci, 14/e, pp 2106–2108.)* One of the least recognized causes of infertility in a woman is scarring of the uterus postpartum: Asherman's syndrome. It classically follows curettage of the uterus, such as occurred here. Women with this syndrome are infertile because of an inability to implant. Ovarian failure, hypothyroidism, and prolactinoma are all eliminated because she still has scant regular menses. Endometriosis causes painful menses.

462. The answer is b. *(McPhee, 2/e, pp 539–540; Barron, 2/e, pp 4, 10, and 285–286.)* HELLP syndrome is hemolysis, elevated liver enzymes, and low platelets (thrombocytopenia). Hypertension is associated with HELLP but is not a part of the syndrome.

463. The answer is d. (*McPhee, 2/e, pp 454, 520–521, and 544–546; Fauci, 14/e, pp 1983–1984.*) LH causes the thecal cells to produce androgens, granulosa cells to produce progesterone, and Leydig cells to produce androgens. The corpus luteum is stimulated by hCG, once a woman is pregnant. Ovulation is stimulated by both LH and FSH.

464. The answer is d. (*McPhee, 2/e, pp 454, 520–521, and 544–546; Fauci, 14/e, pp 1984.*) FSH causes ovarian follicles to mature, granulosa cells to activate aromatase, ovulation to occur along with LH, and Sertoli cells to produce testicular fluid for transporting sperm. The Leydig cells are stimulated by LH.

465. The answer is a. (*McPhee, 2/e, pp 432, 436, and 529; Barron, 2/e, pp 74–75.*) This patient has the classic triad of diabetes mellitus—polyuria, polyphagia, and polydipsia—in combination with visual problems, which can be a marker of diabetic retinopathy. Deep venous thrombosis, urinary tract infection, and preeclampsia are all complications of pregnancy, but none present like this.

466. The answer is c. (*McPhee, 2/e, pp 526–527.*) Although all five options sound appealing, only protection of the baby from infection is a proven benefit. This occurs through immunoglobulins (IgA) in the breast milk. Now that various formulas exist, nutrition no longer is a major reason by itself to breast feed. Quicker weight loss is an old wives' tale. The reliability of breast feeding as a contraceptive technique is low, at best.

467. The answer is e. (*McPhee, 2/e, pp 455–456 and 526–528; Fauci, 14/e, pp 1974–1978 and 2115–2116.*) Prolactin stimulates milk production. Oxytocin stimulates milk ejection. Good maternal nutrition is needed to assure the adequate nutritional content of the milk. An intact neurologic axis is needed to assure that the sucking by the baby leads to oxytocin and prolactin secretion. High estrogen levels inhibit milk production.

468. The answer is e. (*McPhee, 2/e, pp 461–462 and 549; Fauci, 14/e, p 2092.*) FSH assures sperm production. LH assures androgen production. Testosterone is the end product of the testes, and deficiency causes infertility. Cortisol deficiency stimulates an increase in prolactin secretion. High prolactin levels are a cause of male infertility.

148 Pathophysiology

469. The answer is b. *(McPhee, 2/e, pp 542 and 548–549; Fauci, 14/e, p 2093.)* Varicoceles increase the temperature of the scrotum, and hence the testicles, by an increase in local blood flow. Sperm production is reduced by high temperatures. Testosterone production and testicular atrophy are not caused by varicoceles.

470. The answer is e. *(McPhee, 2/e, p 522; Fauci, 14/e, pp 2110–2113.)* Nausea, blood clots, and an increased risk of some cancers are undesirable side effects of oral contraceptives. Contraception is created by blocking the midcycle FSH/LH surge and changing the cervical mucus and endometrial lining. These medications cannot protect against infections.

471. The answer is e. *(McPhee, 2/e, pp 481 and 538; Fauci, 14/e, 2114.)* Pap smear may detect a malignancy as the cause of the bleeding. Hypothyroidism is also a cause of postmenopausal bleeding. A bleeding disorder, as documented by an elevated prothrombin time, can explain bleeding. Vaginal infections can also cause vaginal bleeding. A general chemistry evaluation at this time is too nonspecific.

472–476. The answers are a, a, b, d, c. *(McPhee, 2/e, p 516; Fauci, 14/e, pp 2098–2100.)* Müllerian-inhibiting substance, plasminogen activator, activin, inhibin, follicle regulatory protein, insulin-like growth factor-1, epidermal growth factor-like, platelet-derived growth factor, pro-opiomelanocortin, and gonadotropin surge-inhibiting factor are produced in the granulosa. Renin, inhibin, relaxin, and transforming growth factors α and β are produced in the theca. Basic fibroblast growth factor is produced in the corpus luteum.

477–481. The answers are a, b, b, a, e. *(McPhee, 2/e, pp 548–549; Fauci, 14/e, pp 2092–2095.)* Pretesticular causes are those that affect the hormones that stimulate the testicles, such as a low LH or FSH level. Phenytoin acts by reducing FSH. Testicular causes are those with a direct effect on the testicles, such as alcohol by causing atrophy, and mumps by direct infection. Posttesticular causes are those that affect sperm transport, such as hypospadias.

Nervous System

Questions

DIRECTIONS: Each item below contains a question or incomplete statement followed by suggested responses. Select the **one best** response or the **matching** response to each question.

482. A 35-year-old man presents to you, accompanied by his wife, because he has recently had headaches, which are generalized in nature. His wife reports that he has been somewhat confused at times and clumsy. He does not confirm this but does report that he has been short of breath at times. On your exam, he is noted to be mildly tachycardic and there is a subtle, reddish appearance to his mucous membranes. The only recent new medical problem identified is that the patient was in a motor vehicle accident in which his vehicle was struck in the rear end and it left him with a sore neck. The most likely diagnosis is

a. acquired spinal stenosis (cervical)
b. normal pressure hydrocephalus
c. carbon monoxide poisoning
d. cocaine toxicity
e. muscle tension

483. A patient is noted to have miosis of the right eye and ipsilateral ptosis and reports that he has noted that that side of his face was not sweating when he was working recently. The most likely cause of this clinical picture may be

a. Pancoast's tumor
b. brainstem cerebrovascular accident (CVA)
c. dissection of the carotid artery
d. idiopathic

484. A 55-year-old man describes bilateral pain in his lower back and legs upon prolonged standing while working on an assembly line. Whenever he sits down and takes a break, he gets some relief, but then it recurs when he resumes his job. No other inciting events can be identified. The most likely cause of this problem is

a. degenerative joint disease
b. degenerative disc disease
c. peripheral vascular disease
d. lumbar spinal stenosis

485. An elderly woman has had loss of vision in her left eye. This loss had been transient on a couple of occasions but is now persisting. She has been seen recently at urgent care centers for multiple complaints, including generalized fatigue, left-sided dull boring headaches with occasional sharp jabbing sensations, and arthritic complaint in the hips. Additionally, she reports a recent loss of 7 to 10 pounds. The only remarkable finding on the routine lab tests obtained from her prior evaluations is an elevated alkaline phosphatase. You determine that the likely cause of her condition is
a. glaucoma
b. brain tumor arising anterior to the optic chiasm
c. optic neuritis
d. temporal arteritis

486. A middle-aged woman arrives at your office after not seeing a physician for many years to have a general physical done. Overall, she appears healthy, but a poorly reacting left pupil is noted when a pen light is used. Instillation of a weak solution of pilocarpine leads to constriction of the pupil quickly. Your diagnosis is
a. Argyll-Robertson pupil
b. Adie's tonic pupil
c. trauma-induced pupil dysfunction
d. Horner's syndrome

487. In the aforementioned patient, what other abnormality on physical exam would be consistent with the diagnosis?
a. Ptosis
b. Decreased visual acuity on the Snellen chart
c. Hyporeflexia in the lower extremities
d. Ataxic gait

488. While in the intensive care unit, you are called to your patient's bedside because of the development of seizure activity in this ventilator patient with nosocomial pneumonia. You review the situation, including the medication record. The patient is currently on dopamine, one-half normal saline, imipenin/cilistatin, tobramycin, lisinopril, clonidine patch, and famotidine. The lab results from this morning show normal electrolytes except for a mildly elevated creatinine of 2.4 μg/dL, which is chronic, and the CBC shows an improving white blood count of 15,000 mm^3. After stopping the acute seizure event, you determine the next step in preventing further seizures is
a. stop dopamine
b. stop clonidine
c. phenytoin loading IV
d. change antibiotic coverage
e. CT of the head

489. Two days after admission, a 57-year-old man suddenly has a seizure. He was undergoing evaluation for substernal chest pain. He had been noted by the nursing staff to be a little shaky since shortly after arrival on the floor. He has no history of seizures and the results of a stat CT of the head done during the postictal state were normal. Lab tests had shown an elevated mean corpuscular volume (MCV) of 101, and his aspartate aminotransferase (AST) level was mildly elevated (1½ normal). In this patient, the most likely imbalance that would contribute to this event would be

a. folate deficiency
b. fasting hypoglycemia
c. hypomagnesemia
d. thiamine deficiency
e. vitamin B_{12} deficiency

Items 490–493

Match the one best answer from this list with the following descriptions:

a. Parkinson's disease
b. Essential tremor
c. Asterixis
d. Hyperthyroidism
e. Drug related

490. High-frequency fine tremor that may be difficult to see grossly

491. Resting tremor that lessens with intentional movement

492. Associated with hepatic encephalopathy

493. Lessens with consumption of small amounts of alcohol

494. A 65-year-old woman is seen for evaluation of dementia. On her exam, you note the presence of a left pupil that does not react well to light. While her eyes are following your finger, as you approach the bridge of her nose, you note the pupil to constrict equally as well as the right one. The most important test to order at this point would be

a. Lyme titer
b. vitamin B_{12} level
c. rapid plasma reagent (RPR)
d. HIV
e. fasting glucose

495. Muscle weakness can be demonstrated in patients with all of the following disorders EXCEPT
a. Lambert-Eaton syndrome
b. side effects due to aminoglycosides
c. botulism
d. cerebellar degeneration
e. myasthenia gravis

496. A 68-year-old man has right-sided jaw pain that occurs about halfway through his meal. He has seen his dentist already, but no abnormality was found. X-rays of the area were also taken then but were unremarkable. The next step in his workup should be
a. CT of the region
b. CBC
c. erythrocyte sedimentation rate (ESR)
d. calcium level
e. carotid duplex

497. Which is TRUE regarding the optic neuropathy that occurs with combined use of cigarettes and alcohol?
a. It is reversible with vitamin B supplements and abstinence.
b. Color vision is not impaired.
c. It is of sudden onset.
d. Scotomas are present centrally.

498. Entrapment of the median nerve, causing carpal tunnel syndrome, and, if severe, muscle wasting in the thenar eminence of the hand, is associated with which of the following diseases?
a. Hypothyroidism
b. Diabetes
c. Amyloidosis
d. Rheumatoid arthritis
e. All of these

499. A 75-year-old man with concerns over developing dementia problems is brought to your office by his son. Previously, the patient had been well but was forced to retire from his job a few months ago due to worsening arthritis symptoms limiting his mobility. He has been a widower for 4 months and lives alone. His family is worried about his safety in view of these changes. The likely cause of this dementia picture is
a. HIV related
b. vitamin B_{12} deficiency
c. depression
d. multi-infarct dementia

500. In myasthenia gravis, the antibody to the acetylcholine receptor acts via all the following mechanisms EXCEPT
a. destruction of postsynaptic receptors
b. reduction in the presynaptic release of acetylcholine with repetitive stimulation
c. damage to the muscle membrane on the postsynaptic muscle
d. binding to the receptor site

501. A 38-year-old man presents to the emergency room after the sudden onset of a severe headache while chopping wood. This is the worst pain he has ever experienced, and it is accompanied by photophobia. Evidence of the underlying disease process may be found on
a. CT of the head
b. lumbar puncture
c. ECG
d. all of these

502. Excitatory neurotransmitters such as acetylcholine and glutamate
a. open cation channels and allow influx of Na^+ or Ca^{++}
b. generate inhibitory postsynaptic potentials
c. activate mitochondria
d. regulate intracellular K^+
e. carry impulses between peripheral nerves only

503. A 25-year-old graduate student is injured in a fall during a weekend rock-climbing expedition. There is serious damage to the peripheral nerves in his lower leg. All of the following can be expected to occur EXCEPT
a. Denervated muscles will atrophy.
b. Individual muscle fibers may contract spontaneously and be detected by electromyography (EMG) as fibrillation.
c. Groups of muscle fibers may spontaneously discharge and cause visible fasciculation.
d. Normal motor function will return within 4 weeks.
e. Muscle bulk may decrease by one-half within 2 to 3 months.

504. Weakness is caused by impaired neuromuscular transmission in all of the following disorders EXCEPT
a. myasthenia gravis
b. botulism
c. amyotrophic lateral sclerosis (ALS)
d. aminoglycoside antibiotic-associated weakness
e. Lambert-Eaton myasthenic syndrome

505. Cerebellar disease causes all the following clinical symptoms EXCEPT
a. hypotonia
b. ataxia
c. paralysis
d. dysmetria
e. intention tremor

Items 506–510

Match the symptom description to the nerve disorder:
a. Mononeuropathy
b. Brown-Séquard syndrome
c. Polyneuropathy
d. Mononeuropathy multiplex
e. Radiculopathy

506. Pain and paresthesias (such as a burning or tingling sensation) in both feet and lower legs of a patient who has been diabetic for 15 years

507. Pain and paresthesia over the lateral/anterior thigh of a moderately obese factory worker who stands for long periods of time

508. Impaired pain and temperature sensation in one leg and impaired proprioception and vibration sense in the opposite leg of a patient who was stabbed in the back during a fight

509. Pain and paresthesia in scattered regions of both arms and legs in a patient known to have vasculitis associated with rheumatoid arthritis

510. Pain and weakness in the lower back, radiating down the posterolateral thigh and lower leg of a patient who has worked for many years as a stevedore

511. An amenorrheic 35-year-old woman with galactorrhea is found to have a large prolactin-secreting pituitary tumor compressing her optic chiasm. Which visual disturbance does she have?
a. Left central scotoma
b. Bitemporal hemianopia
c. Left nasal hemianopia
d. Left homonymous hemianopia
e. Completely blind left eye

Items 512–515

Match the hearing disorder to its pathophysiologic description:
a. Conductive deafness
b. Sensorineural deafness
c. Central deafness
d. Tinnitus

512. Deafness caused by disease of the cochlear portion of the eighth cranial nerve

513. Subjective sensation of noise in the ear

514. Deafness due to disease of the cochlear nuclei or auditory pathways

515. Deafness due to disease of the external or middle ear

516. Failure of arousal (*coma*) may be caused by all the following EXCEPT
a. cerebral hemorrhage, either subarachnoid or intracerebral
b. large strokes caused by cortical infarction
c. Creutzfeld-Jakob disease or multi-infarct dementia
d. ethanol or opiate poisoning
e. metabolic encephalopathies, such as hepatic encephalopathy

517. Patients commonly complain of *dizziness*. This symptom is
a. a well-defined sensation of one's environment spinning around
b. usually accompanied by the physical finding of nystagmus
c. usually accompanied by hearing loss
d. a loosely defined and nonspecific symptom of light-headedness or weakness or spinning
e. A specific sign of impending stroke

Items 518–522

Match the disorder to its definition:
a. Apraxia
b. Aphasia
c. Abulia
d. Anomia
e. Alexia

518. The inability to understand speech or to perform meaningful speech

519. The inability to retrieve from memory or to use previously learned words appropriately

520. A blunted and apathetic affect caused by frontal lobe damage

521. The inability to read printed words

522. The inability to perform previously learned motor functions

523. All of the following have been identified as factors in the pathophysiology of amyotrophic lateral sclerosis (ALS) EXCEPT
a. neurofilament protein abnormalities
b. Lewy bodies
c. glutamate transport abnormalities
d. damage from free radicals
e. abnormalities in the SOD gene on chromosome 21

524. Tonic–clonic seizures are characterized by all of the following EXCEPT
a. three per second (3 Hz) spike and wave activity on electroencephalogram (EEG)
b. sudden loss of consciousness
c. a 10- to 30-s phase of tonic muscle contraction with arching of the back
d. postictal confusion usually lasting a few minutes
e. a clonic phase of limb jerking that usually lasts from 30 to 60 s

525. Which clinical scenario best describes a patient with midstage dementia of the Alzheimer's type?
a. A patient has gradually developed memory deficits over the past 4 or 5 years. The deficits worsen each time he has a "spell," described by the family as "little strokes that get better in a few days."
b. A patient has had Parkinson's disease for years and, after becoming nearly immobile, he is also noted to have memory and language deficits.
c. A patient has become progressively more withdrawn and shows deficits in short-term and long-term memory. These deficits have been noticed by the family since the patient's wife and his last sibling died about 5 months ago.
d. A patient became socially withdrawn a couple of years ago because he could not keep up with his friends' activities, such as golf and bridge. Now he is getting lost whenever he leaves his house.
e. A long-term, often homeless, alcoholic becomes progressively disoriented and confused, and the condition cannot be reversed by a move to a nursing home where he receives adequate nutrition and medical care.

526. Which cerebral artery is blocked in an ischemic stroke that presents with the following symptoms: aphasia, right hemiparesis, and right arm numbness?
a. Right anterior cerebral
b. Right middle cerebral
c. Right proximal posterior cerebral
d. Left anterior cerebral
e. Left middle cerebral

Nervous System

Answers

482. The answer is c. (*Fauci, 14/e, p 2533.*) This is the classic presentation of carbon monoxide poisoning. There would be a discrepancy between the pO_2 and O_2 saturation on the arterial blood gas (ABG) evaluation. Pulse oximetry would be correct in the O_2 saturation estimation.

483. The answer is d. (*Fauci, 14/e, pp 559 and 160; Adams, 5/e, p 470.*) Although all of these processes may cause Horner's syndrome, the most common cause remains idiopathic. The Pancoast tumor and dissection of the carotid impinge on the sympathetics, thus exerting their effect. Brainstem strokes would work at the central level to interrupt the sympathetics.

484. The answer is d. (*Fauci, 14/e, p 77.*) This syndrome is called pseudoclaudication. It may also occur at times with exertion, thus causing confusion with peripheral vascular disease. Nerve impingement by osteoarthritis and by degenerative disc disease tends to give a radicular pattern to the discomfort.

485. The answer is d. (*Fauci, 14/e, pp 71 and 1917–1918.*) There is a high correlation between temporal arteritis and the occurrence of polymyalgia rheumatica, and this would explain the proximal muscle girdle pain that is frequently found in temporal arteritis. This disease is an inflammation of the small arteries, although there may be some involvement of the middle-sized arteries. The only lab test to use in an attempt to confirm your diagnosis is to obtain an ESR, which should be elevated above 100 mm/h. The definitive diagnosis is by temporal artery biopsy.

486. The answer is b. (*Fauci, 14/e, p 160.*) This disorder is benign and generally noted in younger females, where it is felt to represent a mild dysautonomia. Tonic pupils may also be seen in diabetes, segmental hypohydrosis, Shy-Drager syndrome, and amyloidosis.

158 Pathophysiology

487. The answer is c. *(Fauci, 14/e, p 160.)* Hyporeflexia in the lower extremities may be seen with this benign condition.

488. The answer is d. *(Fauci, 14/e, p 865.)* β-Lactam antibiotics, in particular high-dose penicillin G and imipenin, are known to induce seizures especially in the face of renal dysfunction. Acute treatment of the seizure would be the same as for any other source of seizure. Other medications could contribute to lowering seizure threshold via lowering the magnesium level (such as with diuretics). The use of phenytoin should not be necessary unless recurrent events occur. The CT of the head is reasonable, but discontinuation of the β-lactam would be the first step.

489. The answer is c. *(Fauci, 14/e, pp 2504–2505 and 2265.)* The consumption of alcohol leads to excessive loss of magnesium in the urine and thus lowers seizure threshold. The AST level in general will elevate more in the face of alcohol usage than will the ALT level. The bulk of alcohol withdrawal seizures will occur within 5 days of the cessation of alcohol consumption.

490–493. The answers are d, a, c, b. *(Fauci, 14/e, pp 113–114, 424t, and 1716.)* The tremor of hyperthyroidism is very high frequency and fine, unlike the other tremors, and does not depend upon rest or movement. The tremor of Parkinson's disease is a coarse resting tremor that decreases with intentional movement. Cogwheel rigidity is also seen. In hepatic encephalopathy, the "flapping" tremor of asterixis is seen. Drug-related tremors are usually seen with $β_2$ agonists and methylxanthines. These are frequently of abrupt onset and time related to the usage or dosage adjustment with these medications.

494. The answer is c. *(Fauci, 14/e, p 160.)* This is the typical Argyll-Robertson pupil found in syphilis.

495. The answer is d. *(Fauci, 14/e, pp 2469–2472.)* The Lambert-Eaton myasthenic syndrome is associated with malignancy, especially small cell lung cancer. Aminoglycosides and penicillamine may cause this picture in normal individuals or exacerbate the disease in patients with myasthenia. In botulism, toxin production mimics myasthenia.

496. The answer is c. *(Fauci, 14/e, pp 1917–1918.)* Jaw claudication is a classic presentation of temporal arteritis. The ESR in this disease should be greater than 100 mm/h in general. The definitive diagnosis is by temporal artery biopsy only.

497. The answer is d. (*Fauci, 14/e, p 2457.*) The typical deficiency amblyopia that occurs with these substances is a gradual process with no reversibility, so only stability of the current status can be achieved. Color vision may be affected, and scotomas in the central or paracentral location are common.

498. The answer is e. (*Fauci, 14/e, pp 1933–1934 and 2466.* Carpal tunnel syndrome is typically associated with repetitive use of the hand/forearm, and multiple diseases may contribute also, including all of these. Pregnancy may also cause this condition to surface; in this situation, however, it generally improves postpartum.

499. The answer is c. (*McPhee, 2/e, p 156; Fauci, 14/e, p 147.*) As many as 10 to 15 percent of patients evaluated for dementia are found to have depression. Care must be taken not to overlook this diagnosis as the underlying cause of dementia or as an aggravating factor.

500. The answer is b. (*Fauci, 14/e, pp 2469–2472.*) The antibody does not cause a reduction in the amount of presynaptic acetylcholine being released, because this is a natural phenomenon. It does bind the receptor site and cause the destruction of receptors and, via the activation of complement, damage the muscle membrane.

501. The answer is d. (*Fauci, 14/e, p 2345.*) Subarachnoid hemorrhage (SAH) is associated with cocaine use, berry aneurysm, AV malformations, and extension of primary intracerebral hemorrhage. It is also reported to occur in conjunction with exertion. Episodes of SAH are typically of sudden onset and severe in nature. The lack of objective neurologic findings is common. The severity of the episode and the presence of vomiting along with loss of consciousness 50 percent of the time suggests this diagnosis. The CT will miss 20 percent of these; thus, lumbar puncture is required to exclude this diagnosis completely. The ECG may have a prolongation of the QRS complex, prolongation of the QT interval, or T-wave changes consisting of inversions or tall peaked waves.

502. The answer is a. (*McPhee, 2/e, p 126.*) Excitatory neurotransmitters generate excitatory postsynaptic potentials by opening channels that allow Na^+ and Ca^{++} to enter neurons. Different neurotransmitters, such as GABA and glycine, cause inhibition of signals.

503. The answer is d. (*McPhee, 2/e, p 131.*) Fasciculation, fibrillation, and substantial atrophy can be expected. Normal muscle function will not likely return without aggressive medical intervention and may not return at all.

504. The answer is c. (*McPhee, 2/e, pp 130 and 149; Fauci, 14/e, pp 2368 and 2469–2472.*) Impaired function of (aminoglycosides) or antibodies to (Lambert-Eaton) calcium channels in nerve terminals, toxins that prevent neurotransmitter release (botulism), and antibodies to neurotransmitter receptors (myasthenia gravis) all cause weakness in different clinical syndromes. The weakness in ALS is due to diseased motor neurons in the anterior horn cells.

505. The answer is c. (*McPhee, 2/e, p 134; Fauci, 14/e, p 116.*) The cerebellum functions as a coordinating center for the maintenance of muscle tone and the regulation of motor tasks, rather than supplying nerve impulses to cause muscles directly to contract.

506–510. The answers are c, a, b, d, e. (*McPhee, 2/e, p140; Fauci, 14/e, pp 2460–2468.*) Numerous localized disorders and many systemic diseases can damage the spinal cord or the peripheral nerves. The pattern of pain, sensory loss, and sometimes weakness can help to classify the disorder.

511. The answer is b. (*McPhee, 2/e, p 142; Fauci, 14/e, pp 161–163 and 1974–1975.*)

512–515. The answers are b, d, c, a. (*McPhee, 2/e, p 145; Fauci, 14/e, pp 175–179.*) Sounds must be conducted through the middle ear and sensed by the cochlea and cranial nerve (CN) VIII; then they are processed by the cochlear nuclei and CNS pathways. Tinnitus is a very annoying, but usually benign, problem often caused by the cochlear portion of the eighth nerve disorders. Hearing may be diminished, but patients are not rendered fully deaf.

516. The answer is c. (*McPhee, 2/e, pp 147–148; Fauci, 14/e, pp 125–134.*) Many toxins and metabolic disturbances cause coma, as do structural lesions, such as hemorrhages and large infarcts. Dementing illnesses do not ordinarily cause coma although, in their terminal stages, patients may be bedridden and virtually unresponsive.

517. The answer is d. (*McPhee, 2/e, p 146; Fauci, 14/e, pp 100–107.*) Dizziness is a nonspecific term and an ill-defined symptom sometimes describing faintness or weakness and sometimes meaning true vertigo. *Vertigo* should be used only when there is an actual sensation of movement; usually, the patient feels that he or she is spinning or the room is spinning around them.

518–522. The answers are b, d, c, e, a. (*McPhee, 2/e, p 148; Fauci, 14/e, pp 134–142.*) Although various areas of the cerebral cortex are specialized to perform certain functions, the cognitive and behavioral domains (e.g., language, memory, and calculation ability) are interconnected to both cortical and subcortical neural networks. Specific deficits identified in a careful neurologic exam will help to define which area or network has been damaged.

523. The answer is b. (*McPhee, 2/e, pp 150–151.*) Lewy bodies are associated with Parkinson's disease. The other abnormalities have been linked to ALS, though its precise pathophysiology is still undefined.

524. The answer is a. (*McPhee, 2/e, pp 154–155; Fauci, 14/e, pp 2311–2324.*) The EEG abnormality described is typical of *absence* seizures. The other characteristics describe a typical tonic–clonic (or *grand mal*) seizure.

525. The answer is d. (*McPhee, 2/e, pp 155–158; Fauci, 14/e, pp 2348–2353.*) Scenario A describes a patient with probable multi-infarct dementia; scenario B is a patient with *Parkinson's–dementia complex* or perhaps Lewy body disease; scenario C is consistent with the pseudodementia of depression. Scenario D is the description most closely associated with Alzheimer's disease, though students should realize that the clinical syndromes may overlap considerably. Scenario E describes a long-term alcoholic whose dementia may be multifactorial, with possible direct toxic damage to the brain, possible nutritional deficits, and a social history that might well include episodes of head trauma.

526. The answer is e. (*McPhee, 2/e, p 160; Fauci, 14/e, pp 2328–2336.*) An infarction of the left hemisphere causes weakness or paralysis and sensory loss on the right. Most right-handed patients have their dominant speech center on the left, and it is supplied by the middle cerebral artery, as is the somatic motor area.

High-Yield Facts

HIGH-YIELD FACTS FOR PATHOPHYSIOLOGY

1. Many diseases have an immunologic basis. Example: **DiGeorge syndrome** (congenital thymic aplasia) is caused by embryologic neural crest development abnormalities affecting the thymus, cardiovascular system, and parathyroid. A congenitally small thymus with decreased immune function causes an isolated T cell deficiency and greater susceptibility to infections by mycobacteria, fungi, or viruses. In severe disease, children often are born with cardiac and facial birth defects. (McPhee, 2/e, pp 40–41.)
2. The following chart compares **bacterial meningitis** and **viral meningitis**. (McPhee, 2/e, pp 61–63.)

	Bacterial Meningitis	Viral Meningitis
Disease state	Acute: significant mortality without antibiotic therapy	Acute: usually self-limited
Symptoms	Fever Worst headache of life Meningismus Mental status changes	Fever Worst headache of life Meningismus Mental status changes
Physical exam findings	Photophobia Nausea Vomiting Fever Kernig's sign—positive Brudinski's sign—positive	Photophobia Nausea Vomiting Fever Kernig's sign—positive Brudinski's sign—positive
Etiology	Neonates *Escherichia coli* Group B *Streptococcus* *Listeria monocytogenes* Children *Neisseria meningitidis* *Streptococcus pneumoniae* *Haemophilus influenzae*, nonimmunized	Cocksackie A and B viruses Poliovirus Mumps virus Epstein-Barr virus Adenovirus Cytomegalovirus

	Bacterial Meningitis	**Viral Meningitis**
Etiology (con't)	Adults (more than 18 years old) N. meningitidis S. pneumoniae L. monocytogenes Gram-negative bacilli	
Cerebrospinal fluid results	Decreased glucose Increased protein Increased neutrophils Increased pressure Gram stain shows bacteria	Normal glucose Slightly increased protein Increased monocytes Normal or slightly increased pressure Gram stain shows no bacteria
Treatment	IV antibiotics Supportive therapy	Supportive therapy
Complications	Cerebral edema Deafness Death	Deafness Weakness

3. Carcinomas undergo phenotypic transition from **normal → hyperplasia → carcinoma in situ → invasive carcinoma → metastasis**. Carcinomas occur as a result of a constellation of physiologic and genetic changes (for example, APC, hMLH1, and hMSH2—colon carcinoma/BRCA1 and BRCA2—breast carcinoma). (McPhee, 2/e, pp 83–84.)

4. **Colon carcinoma** begins when cell cycle regulation loses control over growth, and a collection of rapidly multiplying cells (**hyperplasia**) form an adenoma. The adenoma can continue to develop into **carcinoma in situ**. The first evidence of disease may be occult rectal bleeding indicating the appearance of new friable vessels supplying the tumor. Next, the cancer cells invade the basement membrane of the colon (**invasive carcinoma**), gaining access to the body's transport systems (lymphatic and hematogenous). **Metastasis** to lymph nodes and distant body regions can occur. (McPhee, 2/e, pp 85–87.)

5. Many malignancies have characteristic indirect systemic effects via multiple mechanisms. In lung malignancies, excess adrenocorticotropic hormone (ACTH) production results in a Cushing-like syndrome and excess antidiuretic hormone (ADH) production results in a syndrome of inappropriate antidiuretic hormone secretion (SIADH).

Malignancies (e.g., breast carcinoma) can produce PTH (or related peptides), causing hypercalcemia. Carcinoid syndromes produce serotonin prostaglandins that can cause flushing, restrictive lung symptoms, ascites, and hypotension. (McPhee, 2/e, p 96.)

6. **Pernicious anemia** is a common cause of chronic vitamin B_{12} deficiency. Antibodies to intrinsic factor and parietal cells attack the gastric mucosa, causing gastric atrophy. The disruption of the normal function of the gastric mucosa affects vitamin B_{12} absorption on two levels: stomach acid deficiency (achlorhydria) prevents the release of vitamin B_{12} from food digestion, and intrinsic factor is necessary for vitamin B_{12} absorption in the terminal ileum. The chronic loss of vitamin B_{12} results in abnormal RBC maturation without changes in hemoglobin synthesis. (Fauci, 14/e, pp 655–656; McPhee, 2/e, p 111.)

7. Pathophysiology of hearing loss (McPhee, 2/e, pp 145–146.)

Type of Hearing Loss	Etiology	Testing
Conductive deafness	Disruption of conduction and amplification of sound from the external auditory canal to the inner ear	Negative Rinne test Weber test: heard best in the affected ear Audiometry
Sensorineural deafness	Impaired function of inner ear or cranial nerve VIII	Positive Rinne test Weber test: heard best in the unaffected ear Audiometry
Central deafness	Damaged CNS auditory pathways	Audiometry

8. **Myasthenia gravis** is an autoimmune disease characterized by a deficiency in the number of acetylcholine receptors on the postsynaptic (muscle) terminal, resulting in reduced efficiency of neuromuscular activity. The disease commonly presents in small muscle groups, accompanied by intermittent fatigue and weakness relieved by rest. (McPhee, 2/e, pp 152–153.)

9. **Psoriasis** is an inflammatory parakeratotic accumulation of skin cells that features erythematous, demarcated lesions with scaly patches commonly found on scalp, extensor surfaces of extremities, and fingernails. (McPhee, 2/e, pp 169–170.)

168   Pathophysiology

10. **Asthma** is an obstructive pulmonary disease characterized by airway narrowing as a result of smooth muscle spasms, inflammation, edema, and thick mucus production. The pathophysiologic response is mediated by local cellular injury, lymphocyte activation (antigen exposure, B cell activation, and cytokine activity), IgE-mediated mast cell (producing histamine, leukotrienes, and platelet-activating factor) and eosinophil activation. (*McPhee, 2/e, pp 200–201.*)

11. Pulmonary function tests: obstructive lung disease versus restrictive lung disease.

Pulmonary Function Test	Obstructive Lung Disease (e.g., Chronic Obstructive Pulmonary Disease)	Restrictive Lung Disease (e.g., Pulmonary Fibrosis)
FVC	↓	↓
FEV_1	↓	↓
$FEV_1\%$	↓	Normal / ↑
TLC	↑	↓
RV	↑	Normal / ↓

FVC, forced vital capacity; FEV_1, forced expiratory volume in 1 s; $FEV_1\%$, FEV_1/FVC; TLC, total lung capacity; and RV, residual volume.

12. **Pulmonary embolism** occurs when a venous thrombi (usually from a deep vein thrombosis) lodges in the pulmonary circulation. The pathophysiology includes hemodynamic changes, increased alveolar dead space with increased ventilation/perfusion ratios, and decreased oxygen perfusion to body tissues. Common acute presentations include tachypnea, tachycardia, fever, cough, and pleuritic pain. (*McPhee, 2/e, pp 214–216.*)

13. In congestive heart failure, systolic dysfunction causes reduced stroke volume and reduced cardiac output. The heart responds by increasing return blood flow to the heart, increasing cardiac output by catecholamine release, and increasing ventricular volume through hypertrophy. (*McPhee, 2/e, pp 230–231.*)

14. **Stable angina** is caused by a fixed partial atherosclerotic plaque in one or more of coronary arteries. When at rest, blood flow is able to

provide adequate oxygenation to the heart muscle. Upon exertion, oxygen demand increases. The partial occlusion prevents adequate oxygenation to the heart, resulting in chest discomfort. Unstable angina is caused by thrombus formation on a fissuring atherosclerotic plaque, which transiently prevents adequate oxygenation to the heart. The resulting ischemia causes chest discomfort whether at rest or during exertion. (McPhee, 2/e, pp 246–248.)

15. Chronic esophageal reflux (as a result of a transient weakened lower esophageal sphincter), alcohol use, and tobacco abuse can result in **Barrett's esophagus**. In the disease, columnar epithelium replaces normal squamous epithelium. Individuals with Barrett's esophagus have an increased risk of developing adenocarcinoma of the esophagus. (McPhee, 2/e, p 306.)

16. **Helicobacter pylori** is a common bacteria that infects the gastric mucosa, providing an increased propensity for peptic ulcer disease through inflammatory mechanisms. Other risk factors for peptic ulcer disease are use of a nonsteroidal anti-inflammatory drug (NSAID), family history, smoking, and Zollinger-Ellison syndrome (gastrinoma). (McPhee, 2/e, p 307.)

17. **Crohn's disease** is a chronic inflammatory bowel disease that affects the whole gastrointestinal tract (from mouth to anus) and is distinguished by alternating regions of normal and full-thickness ulcerations of the bowel wall. Common manifestations are bloody diarrhea, fistula, iritis, arthritis, abscess formation, and small bowel obstruction. (McPhee, 2/e, p 315.)

18. **Ulcerative colitis** is an inflammatory bowel disease that causes continuous, partial-thickness (mucosa only) ulcerations of all or part of the colon and is manifested by bloody diarrhea and abdominal pain. (McPhee, 2/e, pp 315–316.)

19. **Insulin-dependent diabetes mellitus (IDDM)** and **non-insulin dependent diabetes (NIDDM)** differ in multiple ways. IDDM usually starts in young (less than 30 years old), nonobese individuals who sometimes have a family history (weak genetic component). Insulin production deficiency predominates, with rare insulin receptor resistance. IDDM is always treated with exogenous insulin. Diabetic ketoacidosis is a common complication. NIDDM usually starts in older (more than 40 years old), obese individuals who often have a family history of disease (strong genetic component). Insulin receptor

resistance predominates. NIDDM is rarely treated with exogenous insulin. Hyperosmolar coma is a common complication. (McPhee, 2/e, pp 431–445.)

20. **Hyperthyroidism** is characterized by sweating, agitation, weight loss, heat intolerance, palpitations, irritability, and dyspnea. Triiodothyronine (T_3) and thyroxine (T_4) are elevated with concurrent depression of thyroid-stimulating hormone (TSH). **Hypothyroidism** is characterized by fatigue, depression, constipation, weight gain, decreased sweating, cold intolerance, and hoarseness. T_3 and T_4 are depressed with concurrent elevation of TSH. (McPhee, 2/e, pp 475–483.)

21. **Cushing's syndrome** (excessive ACTH production) is characterized by moon facies, neck/trunk obesity, weight gain, mental status changes, purple striae on abdomen, osteoporosis, and glucose intolerance. (McPhee, 2/e, p 497.)

22. **Conn's syndrome** (excessive mineralocorticoid secretion) is characterized by hypokalemia, hypernatremia, hypertension, metabolic alkalosis, glucose intolerance, and weakness. (McPhee, 2/e, p 497.)

23. **Addison's disease** (deficient glucocorticoid secretion) is characterized by weakness, fatigue, weight loss, hypotension, cold intolerance, diarrhea, and anorexia. (McPhee, 2/e, p 497.)

24. **Preeclampsia–eclampsia** is characterized by hypertension, proteinuria, and edema after week 20 of pregnancy. Without treatment, a pattern of complications occurs. Complications include bleeding, malignant hypertension, stroke, renal failure, seizures, disseminated intravascular coagulation, and death. (McPhee, 2/e, pp 539–540.)

25. **Minimal change disease** is the most common cause of nephrotic syndrome in children and is characterized by isolated proteinuria (more than 3.5 g of protein in 24-h urine) and obliterated epithelial podocytes on the glomerular basement membrane. (Fauci, 14/e, pp 1540–1541.)

BIBLIOGRAPHY

Adams RD, Victor M, Ropper AH: *Principles of Neurology*, 5th ed. New York, McGraw-Hill, 1993.
Barron WM, Lindheimer MD: *Medical Disorders During Pregnancy,* 2d ed. St. Louis, CV Mosby, 1995.
Cecil, 20th ed.
Fauci AS, Braunwald E, Isselbacher KJ, Wilson JD, Martin JB, Kasper DL, Hauser SL, Longo DL (eds): *Harrison's Principles of Internal Medicine*, 14th ed. New York, McGraw-Hill, 1998.
Lilly LS (ed): *Pathophysiology of Heart Disease.* Philadelphia, Lea and Febiger, 1993.
McPhee SJ, Lingappa VR, Ganong WF, Lange JD (eds): *Pathophysiology of Disease: An Introduction to Clinical Medicine,* 2d ed. Stamford, CT, Appleton and Lange, 1997.
Murray PR, Rosenthal KS, Kobayashi GS, Pfaller MA: *Medical Microbiology,* 5th ed. St. Louis, CV Mosby, 1997.
Roitt R, Brostoff J, Male D: *Immunology,* 5th ed. London: Mosby International, 1998.

Notes

174 Notes

Notes

Notes

Notes

ISBN 0-07-052692-3

MUFSON: PATHOPHYSIOLOGY